SpringerBriefs in Quantitative Finance

W0080278

SpringerBriefs present concise summaries of cutting-edge research and practical applications across a wide spectrum of fields. Featuring compact volumes of 50 to 125 pages, the series covers a range of content from professional to academic. Briefs are characterized by fast, global electronic dissemination, standard publishing contracts, standardized manuscript preparation and formatting guidelines, and expedited production schedules.

Typical topics might include:

- A timely report of state-of-the art techniques
- A bridge between new research results, as published in journal articles, and a contextual literature review
- A snapshot of a hot or emerging topic
- An in-depth case study

SpringerBriefs in Quantitative Finance showcase topics of current relevance in the field of mathematical finance in a compact format. Published titles will feature both academic-inspired work and more practitioner-oriented material, with a special focus on the application of recent mathematical techniques to finance, including areas such as derivatives pricing and financial engineering, risk measures and risk allocation, risk management and portfolio optimization, computational methods, and statistical modelling of financial data.

Michael Ludkovski • Jimmy Risk

Gaussian Process Models for Quantitative Finance

 Springer

Michael Ludkovski (ID)
Department of Statistics and Applied
Probability
University of California
Santa Barbara, CA, USA

Jimmy Risk
California State Polytechnic University
Pomona, CA, USA

ISSN 2192-7006   ISSN 2192-7014   (electronic)
SpringerBriefs in Quantitative Finance
ISBN 978-3-031-80873-9   ISBN 978-3-031-80874-6   (eBook)
https://doi.org/10.1007/978-3-031-80874-6

This Springer imprint is published by the registered company Springer Nature Switzerland AG
The registered company address is: Gewerbestrasse 11, 6330 Cham, Switzerland

If disposing of this product, please recycle the paper.

Preface

Objective and Audience

This text rigorously integrates Gaussian process (GP) modeling into the fabric of quantitative finance. Spurred by the transformative influence of machine learning frameworks, the volume provides a detailed exposition on the theory and application of GPs. The initial three chapters are tailored to acquaint readers with GP methodology, equipping them with the necessary foundation to navigate the more advanced applications that follow. Targeting both researchers and practitioners, this work is positioned as a reference textbook, offering a broad spectrum of GP model applications that demonstrate their versatility and relevance in quantitative finance.

The book includes online supplementary materials in the form of half a dozen computational notebooks that provide the reader direct illustrations of the covered material. These notebooks (including corresponding datasets and a Readme document) are available to view and download at the Springer Nature GitHub repository: https://github.com/sn-code-inside/gaussian-process-models-for-quantitative-finance.

Guide to the Chapters

Our exposition begins with Chaps. 1 and 2 that form the bedrock of GP theory. The introductory Sect. 1.1 sets the context for GP models within computational finance, highlighting the convergence of machine learning and financial analysis and acquainting the reader with examples that are tackled in later chapters. The rest of Chap. 1 is dedicated to the core methodologies, equipping the reader with the theoretical framework for GP modeling. Chapter 2 goes into details of GP kernels (covariance functions), drawing connections to other mathematical sub-fields, such as reproducing kernel Hilbert spaces and stochastic differential equations. Chapter 3 pivots toward more specialized GP approaches, sampling among non-Gaussian likelihoods, multi-output modeling, and heteroskedastic modeling. This chapter's relevance shifts depending on the specific applications the reader pursues.

The remaining four chapters can be explored independently of one another, each delving into a distinct application of GPs in the financial realm. Specifically,

- Chapter 4 discusses the application of GPs in option pricing and sensitivity analysis.
- Chapter 5 examines Regression Monte Carlo methods, illustrating the use of GPs as functional surrogates for the continuation value in optimal stopping problems.
- Chapter 6 presents the use of GPs for non-parametric modeling of financial objects, such as yield curves, implied volatility surfaces and mortality rate surfaces.
- Chapter 7 addresses the application of GP surrogates in stochastic control.

The brief Appendix offers a review on prerequisite mathematics, like the theory of multivariate Gaussian distributions, linear algebra, and function spaces.

Acknowledgments

We would like to thank the editors at SpringerBriefs in Quantitative Finance for their support, especially Ute McCrory and Remi Lodh for patiently shepherding this project and answering our various questions. We are grateful to the anonymous referees who provided a lot of valuable feedback and went through the manuscript with an astoundingly fine-toothed comb to help us improve the writing. We appreciate the encouragements from series editor Matheus Grasselli when we first proposed this monograph. Last but not least, we extend our gratitude to our peers for their insights and broad support. This book is a testament to the collaborative spirit that drives progress at the intersection of quantitative finance and machine learning.

ML: I would like to thank all my co-authors who greatly contributed to several papers that became parts of this manuscript: Bobby Gramacy, Mickael Binois, Tao Chen, Yuri Saporito, and Howard Zail, as well as my former PhD students (and co-authors) Nhan Huynh, Aditya Maheshwari, and Victoria Lyu. Bobby and Mickael in particular are my personal GP gurus and most of what I know about the topic traces back to them (or their numerous papers on GPs). I would also like to acknowledge various helpful discussions with Rodrigo Targino and Moritz Voss. Most importantly, this project would not have been possible without Jimmy: thank you for being such a wonderful collaborator and your willingness to keep embarking on joint research adventures. Early stages of the writing were in part supported by NSF DMS-1821240.

JR: I would like to thank Mike Ludkovski, my coauthor, colleague, and mentor, for his guidance, patience, and the skills he has taught me throughout the years. I am also grateful to Albert Cohen, my long-time mentor and current colleague, for his support and encouragement both academically and personally. Finally, thank you to Springer for the opportunity to publish this work.

Santa Barbara, CA, USA   Mike Ludkovski
Pomona, CA, USA   Jimmy Risk
October 2024

Contents

Symbols and Notation

Vectors are bolded, as are matrices with data-related vectors. An asterisk, e.g., $\mathbf{x}_*$, indicates reference to a *test set*. The notation $f_*(\cdot)$ or $\mathbf{f}_*$ implies the posterior Gaussian process, i.e., having been conditioned on a *training set* $\mathcal{D}$.

Symbol	Meaning
$\overset{\triangle}{=}$	An equality which acts as a definition
$\lvert \mathbf{K} \rvert$	Determinant of matrix $\mathbf{K}$
$\lvert \mathbf{y} \rvert$	Euclidean length of $\mathbf{y}$ vector: $\left(\sum_j y_j^2 \right)^{1/2}$
$\lvert \mathcal{D} \rvert$	Cardinality of a finite set
$\langle \cdot, \cdot \rangle_{\mathcal{H}}$	Inner product in Hilbert space $\mathcal{H}$
$\lVert \cdot \rVert_{\mathcal{H}}$	Norm induced by inner product $\lVert \cdot \rVert_{\mathcal{H}} = \langle \cdot, \cdot \rangle_{\mathcal{H}}$
$\mathbf{y}^{\top}$	Transpose of vector $\mathbf{y}$
$\sim$	Distributed as; e.g. $\mathbf{y} \sim \mathcal{MVN}(\mathbf{0}, \Sigma)$
∇	Vector of partial derivatives
$\mathbf{0}$ or $\mathbf{0}_n$	Vector of all 0's (of length n)
$\mathbf{1}$ or $\mathbf{1}_n$	Vector of all 1's (of length n)
cov, cor	Covariance and correlation operators
d	Dimension of input space $\mathcal{X} \subseteq \mathbb{R}^d$
$\mathcal{D}$	Data set; $\mathcal{D} = (\mathbf{x}_i, y_i)_{i=1}^N$
δ_{ij}	Kroneker delta, $\delta_{ij} = 1$ if $i = j$ and 0 otherwise
η^2	Outputscale or process variance, determining the unconditional $\mathrm{var}(f(\mathbf{x})) = k(\mathbf{x}, \mathbf{x})$
$\mathbb{E}$	Expected value (expectation) operator
f	Gaussian process, i.e., $f = (f(\mathbf{x}))_{\mathbf{x} \in \mathcal{X}}$
$f(\mathbf{x})$	Gaussian process value at location $\mathbf{x}$
$\mathbf{f}$	Vector of Gaussian process values $\mathbf{f} = [f(\mathbf{x}_1), \ldots, f(\mathbf{x}_n)]^{\top}$
$f_*, f_*(\mathbf{x}), \mathbf{f}_*$	Posterior versions of f, $f(\mathbf{x})$, and $\mathbf{f}$, i.e., conditioned on $\mathcal{D}$

(continued)

Symbol	Meaning
$\mathcal{GP}$	Gaussian process: $f \sim \mathcal{GP}(\mu, k)$ means the process f has mean function μ, covariance function k and multivariate normal finite dimensional distributions
$\mathbf{h}$	Difference of two vectors in $\mathbb{R}^d$, e.g., $\mathbf{h} = \mathbf{x} - \mathbf{x}'$
$\mathcal{H}_k$	Reproducing kernel Hilbert space corresponding to kernel k
$\mathbf{I}$ or $\mathbf{I}_n$	The identity matrix (of size n)
J_ν	Bessel function of the first kind
$k(\mathbf{x}, \mathbf{x}')$	Kernel function evaluated at $\mathbf{x}$ and $\mathbf{x}'$
$k(\mathbf{h})$	Stationary kernel evaluated at $\mathbf{h}$
$k(r)$	Isotropic kernel evaluated at $r = \|\mathbf{h}\|$
$\mathbf{K}, \mathbf{K}_f, K(\mathbf{X}, \mathbf{X})$	$n \times n$ covariance or Gram matrix with i, j entry $k(\mathbf{x}_i, \mathbf{x}_j)$
$\mathbf{K}_*$ or $K(\mathbf{X}, \mathbf{X}_*)$	$n \times n_*$ covariance matrix between training and test cases
$k(\mathbf{x}_*)$ or $\mathbf{k}_*$	Shorthand for vector $K(\mathbf{X}, \mathbf{x}_*)$ for a single test case
$\mathbf{K}_y$	Covariance matrix of the (noisy) $\mathbf{y}$ values $\mathbf{K}_y = \mathbf{K}_f + \boldsymbol{\Sigma}$
$\ell(\boldsymbol{\theta})$	Log-likelihood function
ℓ_{len}	Characteristic length-scale
$\mu(\mathbf{x})$	Prior mean function of a Gaussian process evaluated at $\mathbf{x}$
$m(\mathbf{x}_*)$	Posterior mean function of a Gaussian process evaluated at $\mathbf{x}_*$
$\mathcal{MVN}(\boldsymbol{\mu}, \boldsymbol{\Sigma})$	Multivariate normal with mean vector $\boldsymbol{\mu}$ and covariance matrix $\boldsymbol{\Sigma}$
N and N_*	Number of training (and test) observations
$\mathbb{N}$	Natural numbers, i.e., positive integers
$O(\cdot)$	Big Oh; for functions f and g defined on $\mathbb{N}$, $f(n) = O(g(n))$ if $f(n)/g(n) \leq C$ for some C and all $n \in \mathbb{N}$
$\mathbb{R}$	Real numbers
$r(\mathbf{x})$	Heteroskedastic state-dependence noise variance $r(\mathbf{x}_i) = \text{var}(\varepsilon_i)$
$s(\mathbf{x}_*)^2$	Posterior variance function of a Gaussian process evaluated at $\mathbf{x}_*$ (single argument)
$s(\mathbf{x}_*, \mathbf{x}'_*)$	Posterior covariance function of a Gaussian process evaluated at $\mathbf{x}_*$ and $\mathbf{x}'_*$
σ or $\sigma(\mathbf{x})$	Noise variance
$\boldsymbol{\Sigma}, \boldsymbol{\Sigma}_\epsilon, \boldsymbol{\Sigma}_n$	Covariance matrix of noise vector $\boldsymbol{\epsilon}$
$\boldsymbol{\theta}$	Vector of GP hyperparameters
$\text{tr}(\mathbf{A})$	Trace of (square) matrix $\mathbf{A}$
var	Variance operator
$\Psi(\mathbf{x}, \epsilon)$	Discrete-time transition function of the state process (X_k)
(W_t)	Wiener process / Brownian motion
$\mathcal{X}$	Input space
$\mathbf{X}$	$N \times d$ matrix of training inputs with rows $(\mathbf{x}_i)_{i=1}^N$
$\mathbf{X}_*$	$N_* \times D$ matrix of test inputs
$\mathbf{x}_i$	ith training input
$x_{i,d}$	dth coordinate of $\mathbf{x}_i$
$\mathbb{Z}$	The integers

Chapter 1
Gaussian Process Preliminaries

1.1 Introduction

The advancement of computational finance is increasingly reliant on machine learning frameworks to model, analyze and interpret financial data. Among these methodologies, Gaussian Processes (GPs) emerge as a powerful tool, offering a flexible and transparent framework for financial modeling. This book explores the integration of GPs within the domain of quantitative finance, illustrating their aptitude to address a wide array of challenges faced by practitioners and researchers.

GPs are distinguished by their non-parametric data-driven nature, and by their probabilistic approach. This makes them adept at modeling financial markets, covering the full spectrum of data cleaning, curve fitting, model calibration, sensitivity analysis, functional approximation and more. Some highlights include:

- Derivatives pricing and risk management (Chap. 4). This application showcases the GP's ability to act as a *surrogate model*, for example to learn an input-output relationship through simulation. Compared to traditional models, GPs have built-in ways to (i) naturally incorporate simulation error (arising e.g. through Monte Carlo schemes), and (ii) effectively model the sensitivities of financial instruments (the "Greeks"). Additionally, their probabilistic nature provides uncertainty quantification, a critical requirement in financial risk management.
- An extension of regression Monte Carlo (RMC, also known as the Longstaff Schwartz Algorithm or Least Squares Monte Carlo) framework [53, 107] that replaces timing- and option-value interpolants of American-style options with regularized versions that correspond to GP prediction (Chap. 5). Chapter 7 further extends this concept of GP emulators in value- and policy-iteration to more general stochastic control problems.
- GPs as non-linear curve-fitting models, that can model complex financial objects such as yield curves, volatility surfaces, etc. (Chap. 6).

M. Ludkovski, J. Risk, *Gaussian Process Models for Quantitative Finance*, SpringerBriefs in Quantitative Finance, https://doi.org/10.1007/978-3-031-80874-6_1

The impetus for this volume is the computational revolution that has transformed the numerical toolbox of financial modeling. Rooted in statistical regression and function approximation, machine learning has grown into a vast field spanning everything from Natural Language Processing to Reinforcement Learning and Generative Artificial Intelligence. All these paradigms have been applied to financial problems and the field is poised for further rapid growth, see e.g. the monographs [26, 48]. One core driver has been the success of deep learning—representing latent patterns in (very) high-dimensional spaces through neural network architectures. Driven by efficient new algorithms for gradient descent and large-scale optimization, coupled with successful heuristics for representing high-dimensional objects through function compositions, neural networks (NNs) have been acclaimed as the new gold standard in machine learning. Yet no method can suit all contexts, and various challenges continue to arise in applying deep learning for concrete settings. Gaussian Processes thus offer a useful alternative, in many ways providing a different tack to the machine learning ethos. For example, while NNs thrive in environments with vast (or ideally unlimited) training data, GPs are better suited to sparse training sets. Whereas NNs focus on predictive accuracy, GPs have built-in probabilistic interpretation that organically provides uncertainty quantification in every context. In sum, GPs form a core tool in the machine learning arsenal and present many unique advantages to the modeler.

The goal of this book is to expound the use of GPs in quantitative finance—not to claim their supremacy. We thus structure this monograph as a focused exposé; providing some momentary comparison of how GPs work vis-a-vis other methods (including NNs), but largely staying within the GP ecosystem, prioritizing the elaboration of its nuances and choices. One of the challenges in getting into the GP world-view is that it has been simultaneously shaped and constructed in multiple communities. There is the machine learning community that has led GP efforts in the early 2000s, focusing on prediction tasks, with the most famous application being Bayesian Optimization. There is the applied statistics community, where GPs were originally known as kriging, and continue to be a backbone both of spatial statistics and of Bayesian regression. And there is the applied probability community which could arguably claim to be the original developer of the theory of Gaussian stochastic processes and the respective ideas of conditioning, sample paths, and kernel spaces. All these communities speak their own language, often using a distinct terminology for the same concepts. This makes for a challenging reading for the novice, as well as some duplicative results. Our presentation tries to synthesize all of the above, primarily targeting financial mathematicians who are probably (pun intended) closest to the probabilistic perspective. Hence, we emphasize the conditional equations and point out connections to stochastic differential equations. Otherwise, we freely borrow from a large collection of sources, including Rasmussen and Williams [140], Gramacy [79] plus well over a 100 journal articles. We are also indebted to many excellent online resources that introduce and explain numerous features of GPs, from manuals of software libraries [42, 65, 144] to stand-alone widgets, to lecture notes [51]. For example, see the extensive resources associated with the long-standing Gaussian Process Summer School series (www.gpss.cc).

Applications of GPs in finance is a relatively recent topic, with the vast majority of cited papers appearing after 2015 or so. Hence, our aim is not to provide an authoritative review of a mature subject, but to unify, survey and tie together this emerging topic and its modeling context. Indeed, to our knowledge, so far there is no dedicated monograph on Gaussian processes in quantitative finance. To this end, we have tried to construct a comprehensive bibliography and to outline a variety of GP-related tools, many of them yet to be fully fleshed out in the described applications. Our sincere hope is that this compendium will spur new research, so that eventually a longer and deeper book would be warranted.

The book is organized into two halves. The first half consists of Chaps. 1–3 and provides a cogent introduction and overview of GP modeling theory, serving as requisite knowledge for the later applications, and providing a comprehensive reference to the vast toolbox available for GP modelling. These chapters are structured to progressively guide readers toward the second half consisting of Chaps. 4–7 that focus on different types of applications and can all be independently read/consulted. Specifically, the remainder of Chap. 1 lays the foundation with the theoretical framework which provides predictions and uncertainty quantification, and highlights some connections to existing theory like kernel ridge regression (KRR). Chapter 2 explores the intricacies of GP kernels (covariance functions), connecting them to broader mathematical theories such as reproducing kernel Hilbert spaces (RKHS) and stochastic differential equations (SDEs). This chapter is pivotal for comprehending how GPs can be tailored to model various financial phenomena, like encoding constraints on the underlying input-output relationship. Chapter 3 explores more nuanced situations, for example in incorporating heteroskedastic noise, alternative likelihood functions, multi-output models, and ways to improve computational time through inducing points and variational approximations.

Closing Notes

Each chapter has a section on closing notes, providing references to information that extends the content covered. In addition, reflecting our computational orientation, most chapters include an online supplement, containing code and associated output in a notebook format. The online GitHub repository contains a collection of `Python Jupyter` and R Markdown notebooks that reproduce some of the shown figures and offer a starting point for students learning this material or for researchers who wish to replicate and extend the presented methods.

1.2 Fundamentals

We describe a Gaussian process over an input set X, by default understood to be a subset of $\mathbb{R}^d$. Elements of X are termed inputs, $\mathbf{x} = (x_{(1)}, \ldots, x_{(d)})$. We endow X with the Euclidean metric, in particular we canonically use the difference $|x - x'|$ to measure distances along one coordinate of X.

Definition 1.1 A *Gaussian process* is a collection of random variables $f \triangleq$ $(f(\mathbf{x}))_{\mathbf{x} \in X}$ such that for any finite collection $\mathbf{x}_1, \ldots, \mathbf{x}_N \in X$, the vector $\mathbf{f} \triangleq$ $[f(\mathbf{x}_1), \ldots, f(\mathbf{x}_N)]^\top \in \mathbb{R}^N$ has a multivariate Gaussian distribution.

Note that we purposefully omit the mention of the underlying probability space $(\Omega, \mathcal{F}, \mathbb{P})$; this probabilistic structure is kept in the background so as not to interfere with other sources of uncertainty; for example any randomness in the data. The assumption is that the GP captures aleatoric uncertainty that quantifies model risk, i.e. the intrinsic range of different models that are compatible with the given universe.

Gaussian distributions are characterized through their first two moments, offering a finite-dimensional structure for the entire infinite-dimensional distribution of f. The *mean function* and the *covariance kernel* of a GP are defined as

$$\mu(\mathbf{x}) \triangleq \mathbb{E}[f(\mathbf{x})], \tag{1.1}$$

$$k(\mathbf{x}, \mathbf{x}') \triangleq \operatorname{cov}\left(f(\mathbf{x}), f(\mathbf{x}')\right), \tag{1.2}$$

where for two random variables X_1, X_2, $\operatorname{cov}(X_1, X_2) = \mathbb{E}\left[(X_1 - \mathbb{E}[X_1])(X_2 - \mathbb{E}[X_2])\right]$ is the traditional *covariance operator*. We use the shorthand $f \sim \mathcal{GP}(\mu, k)$ to denote that f is a Gaussian process with mean function μ and covariance kernel k. The mean function determines the overall trend of the process, while its covariance captures the dependencies between different inputs and how these dependencies affect the outputs. Different choices of mean and covariance functions lead to different properties of the corresponding GP, as will be discussed in Chap. 2. Realizations of f are called *sample paths*, reflecting the stochastic process perspective.

It is helpful to think of a Gaussian process as a functional generalization of the Gaussian distribution. The univariate Gaussian distribution starts with scalar mean and standard deviation μ, σ; this is then lifted to a vector μ and a matrix Σ for the multivariate normal distribution $\mathcal{MVN}(\mu, \Sigma)$ and further generalized to a function μ and a kernel k for the GP, see Table 1.1. See Appendix A.2 for a review on the multivariate normal distribution.

Let us analyze the *finite-dimensional distribution (fdd)* of a GP f with a kernel k. Consider $\mathbf{X} = [\mathbf{x}_1, \ldots, \mathbf{x}_N]$, and the $N \times N$ *gram matrix* $\mathbf{K} = K(\mathbf{X}, \mathbf{X}) \triangleq \left[k(\mathbf{x}_i, \mathbf{x}_j)\right]_{i,j=1}^N$, i.e. with i, jth-entry $k(\mathbf{x}_i, \mathbf{x}_j)$. Since this matrix is the covariance

Table 1.1 Comparing univariate Gaussian distribution to multivariate Gaussian, to a Gaussian process

$Y \sim \mathcal{N}(\mu, \sigma^2)$	$[Y_1, \ldots, Y_N]^\top \sim \mathcal{MVN}(\boldsymbol{\mu}, \Sigma)$	$f \sim \mathcal{GP}(\mu, k)$
$\mathcal{X} = \{1\}$	$\mathcal{X} = \{1, \ldots, N\}$	$\mathcal{X} = \mathbb{R}_+, \mathcal{X} = \mathbb{R}^d$, etc.
$\mu \in \mathbb{R}$	$\boldsymbol{\mu} \in \mathbb{R}^N$	$\mu : \mathcal{X} \to \mathbb{R}$
$\sigma^2 \in \mathbb{R}_+$	$\Sigma \in \mathbb{R}^{N \times N}$	$k : \mathcal{X} \times \mathcal{X} \to \mathbb{R}$

matrix of the vector $\mathbf{f}$, it follows that $\mathbf{K}$ is positive semi-definite. The transition from finite index sets to general $\mathcal{X}$ requires the following definition.

Definition 1.2 A function $k : \mathcal{X} \times \mathcal{X} \to \mathbb{R}$ is said to be *positive definite* if, for any finite collection $\mathbf{x}_1, \ldots, \mathbf{x}_N \in \mathcal{X}$, the corresponding matrix $\mathbf{K} = \left[k(\mathbf{x}_i, \mathbf{x}_j)\right]_{i,j=1}^N$ is positive semi-definite.

Remark 1.3 Throughout the book we use $\mathbf{X}$ to denote the $N \times d$ matrix with rows $\mathbf{x}_i, i = 1, \ldots, N$, and for matrices $\mathbf{U}, \mathbf{V}$ of size $\ell \times d$ and $m \times d$ respectively, we use $K(\mathbf{U}, \mathbf{V}) = \left[k(\mathbf{u}_i, \mathbf{v}_j)\right]_{1 \le i \le \ell, 1 \le j \le m}$ to denote the $\ell \times m$ matrix of pairwise covariances.

Thus, the kernel function of any GP is positive definite, which follows from k being defined from pairwise covariances. Similarly, k is a *symmetric* function meaning $k(\mathbf{x}, \mathbf{x}') = k(\mathbf{x}', \mathbf{x})$ for all $\mathbf{x}, \mathbf{x}'$. Henceforth, we use the term *kernel* to refer to a symmetric and positive definite function.

In applications, one typically starts with a mean and covariance functions and then constructs a corresponding GP. Take a measurable function $\mu : \mathcal{X} \to \mathbb{R}$ and a positive definite kernel $k : \mathcal{X} \times \mathcal{X} \to \mathbb{R}$. For any $N \in \mathbb{N}$ and $\mathbf{x}_1, \ldots, \mathbf{x}_N \in \mathcal{X}$, by Definition 1.2 the Gram matrix $\mathbf{K}$ is positive semi-definite. Hence, multivariate normal theory shows that one can construct the vector $\mathbf{f} = [f(\mathbf{x}_1), \ldots, f(\mathbf{x}_N)]^\top$ so that

$$\mathbf{f} \sim \mathcal{MVN}(\boldsymbol{\mu}, \mathbf{K}), \tag{1.3}$$

where $\boldsymbol{\mu} = [\mu(\mathbf{x}_1), \ldots, \mu(\mathbf{x}_n)]^\top$. The Kolmogorov extension theorem [141] then ensures existence of a stochastic process f indexed in $\mathcal{X}$ with mean function μ and covariance function k, with the finite-dimensional distributions of f being multivariate normal (see Appendix A.2). Thus, any such pair (μ, k) satisfies Definition 1.1 to induce a GP f, providing an equivalence between f as a process and its mean-covariance functional pair.

For readers acquainted with probability theory and stochastic processes the above discussion should serve as a review, and indeed look familiar. For example, a GP on $\mathbb{R}^+$ with zero mean and the kernel $k(x, x') = \min(x, x')$ produces the well known *Wiener process* [141]. We now proceed to add a "model" component to connect this foundational theory to applications.

1.3 Gaussian Process Regression

Regression is concerned with the twin tasks of smoothing and interpolating: learning an input-output relationship between $\mathbf{x} \in X \subseteq \mathbb{R}^d$ and the latent output $\mathbf{x} \mapsto f(\mathbf{x}) \in \mathbb{R}$. The response mapping f is not known, and is accessed through some *samples* y's. *Gaussian process regression* is a nonparametric approach to regression problems that utilizes the probabilistic ideas above. Given a *training dataset*

$$\mathcal{D} \triangleq \{(\mathbf{x}_i, y_i) : i = 1, \ldots, N\}, \tag{1.4}$$

the idea is to decompose y_i into an underlying signal $f(\mathbf{x}_i)$ and noise $\epsilon(\mathbf{x}_i)$, whereby the observed $\mathbf{y} \triangleq [y_1, \ldots, y_N]^\top$ provide data-driven inference for $f(\mathbf{x}_*)$ for any $\mathbf{x}_* \in X$ by the conditioning equations. This is formalized as a noisy data generating process $\mathbf{x} \mapsto y(\mathbf{x})$

$$y(\mathbf{x}) = f(\mathbf{x}) + \epsilon(\mathbf{x}). \tag{1.5}$$

We assume $(\epsilon(\mathbf{x}))_{\mathbf{x} \in X}$ is a collection of independent mean-zero random variables with variance $\sigma^2(\mathbf{x}) \geq 0$.

The GP regression assumption is that the underlying "signal" f follows a Gaussian process, that is, $f \sim \mathcal{GP}(\mu, k)$. We think of this GP as the *prior* (with *prior mean* μ and *prior covariance kernel* k), i.e. the default assumption imposed by the modeler before seeing data. The goal of the regression is then to obtain the posterior $f(\mathbf{x}_*)|\mathcal{D}$ for a test point $\mathbf{x}_* \in X$, or more broadly, $f|\mathcal{D}$ (the posterior process). This corresponds to conditioning f on the observed data. Importantly, the probability space that describes sample outcomes of f is *independent* of the observation noise ϵ. In the literature, one distinguishes the former *extrinsic* or aleatoric uncertainty about f from the latter *intrinsic* uncertainty due to ϵ. For now, we assume a zero mean function $\mu(\mathbf{x}) \equiv 0$. The case of $\mu \neq 0$ is discussed in Sect. 1.4.3. A routine computation shows that $\mathbb{E}[y(\mathbf{x})] = \mu(\mathbf{x}) = 0$ and

$$\text{cov}(y(\mathbf{x}_i), y(\mathbf{x}_j)) = k(\mathbf{x}_i, \mathbf{x}_j) + \sigma^2(\mathbf{x}_i)\delta_{ij}. \tag{1.6}$$

Denote by $\mathbf{X}$ the $N \times d$ matrix of stacked training inputs and let $\mathbf{K}_f = [k(\mathbf{x}_i, \mathbf{x}_j)]_{i,j=1}^N = K(\mathbf{X}, \mathbf{X})$ be the covariance matrix of $f(\mathbf{X})$ induced by the kernel k. Furthermore, place the assumption that each $\epsilon(\mathbf{x})$ is normally distributed, $\mathbf{x} \in X$. From Gaussian theory, this gives the fdd $\mathbf{y} \sim \mathcal{MVN}(\mathbf{0}, \mathbf{K}_y)$ where $\mathbf{K}_y = \mathbf{K}_f + \Sigma_\epsilon$ and $\Sigma_\epsilon = \text{diag}(\sigma^2(\mathbf{x}_1), \ldots, \sigma^2(\mathbf{x}_N))$. It follows that the stacked vector $[\mathbf{f}, \mathbf{y}]^\top \sim \mathcal{MVN}$ is multivariate Gaussian, and hence $\mathbf{f}|\mathbf{y} \sim \mathcal{MVN}$. This provides an *in-sample posterior* view of the data.

For a single test point $\mathbf{x}_* \in \mathcal{X}$, a similar argument provides the following predictive distribution. The block covariance matrix of $[f(\mathbf{x}_*), \mathbf{y}]$ is

$$\begin{bmatrix} k(\mathbf{x}_*, \mathbf{x}_*) & \mathbf{k}_*^\top \\ \mathbf{k}_* & \mathbf{K}_y \end{bmatrix} \tag{1.7}$$

where $\mathbf{k}_*^\top \overset{\triangle}{=} K(\mathbf{X}, \mathbf{x}_*)$ is the $N \times 1$ vector of the $k(\mathbf{x}_i, \mathbf{x}_*)$. Then

$$f_*(\mathbf{x}_*) \overset{\triangle}{=} f(\mathbf{x}_*)|\mathcal{D} \sim \mathcal{N}\left(m(\mathbf{x}_*), s^2(\mathbf{x}_*)\right),$$

with

$$m(\mathbf{x}_*) = K(\mathbf{x}_*, \mathbf{X})\left[\mathbf{K}_f + \Sigma_\epsilon\right]^{-1}\mathbf{y}, \tag{1.8}$$

$$s^2(\mathbf{x}_*) = k(\mathbf{x}_*, \mathbf{x}_*) - K(\mathbf{x}_*, \mathbf{X})\left[\mathbf{K}_f + \Sigma_\epsilon\right]^{-1}K(\mathbf{X}, \mathbf{x}_*). \tag{1.9}$$

Similarly, we obtain that $y(\mathbf{x}_*)|\mathcal{D} \sim \mathcal{N}\left(m(\mathbf{x}_*), s^2(\mathbf{x}_*) + \sigma^2(\mathbf{x}_*)\right)$, so the variance of a new observation $y(\mathbf{x}_*)$ is decomposed into the intrinsic randomness $\sigma^2(\mathbf{x}_*)$ and the extrinsic uncertainty $s^2(\mathbf{x}_*)$. Note that much of the machine learning literature exclusively focuses on the latter predictive variance $\mathrm{var}(y(\mathbf{x}_*)|\mathcal{D})$ and omits any mention of $s^2(\mathbf{x}_*)$ itself—a moot point if there is no noise.

Observe that (1.8) is linear in $\mathbf{y}$, that is, the GP regression prediction is a *linear smoother* of the observed data. Indeed,

$$m(\mathbf{x}_*) = K(\mathbf{x}_*, \mathbf{X})\mathbf{K}_y^{-1}\mathbf{y} = \sum_{i=1}^{n} a_i y_i, \tag{1.10}$$

where a_i is the ith entry of $K(\mathbf{x}_*, \mathbf{X})\mathbf{K}_y^{-1}$. Note that the linear coefficients a_i are determined according to the kernel function, and the noise variances. While $\mathbf{K}_y = \mathbf{K}_f + \Sigma_\epsilon$ is present in both (1.8) and (1.9), $\mathbf{y}$ only impact the posterior mean: the posterior variance is a function of the training inputs $\mathbf{X}$ only and is independent of actual observations, solely reflecting the dependence structure induced by the kernel function k and the placement of $\mathbf{x}_i$'s.

Considering (1.9), observe that $s^2(\mathbf{x}_*) \le k(\mathbf{x}_*, \mathbf{x}_*)$, i.e. the posterior variance is smaller than the prior variance $k(\mathbf{x}_*, \mathbf{x}_*)$. This important property of GP regression reflects the idea that when compared with its prior, uncertainty in predictions decreases with the inclusion of training data. This uncertainty reduction is measured by the second term in (1.9) that reflects the contribution of each $\mathbf{x}_i$ via $K(\mathbf{x}_*, \mathbf{x}_i)$. One implication is that when $\mathbf{x}_*$ is "less similar" to the elements in $\mathcal{D}$ (in terms of lower kernel values), the entries in $K(\mathbf{x}_*, \mathbf{X})$ decrease, thereby increasing posterior standard variance $s^2(\mathbf{x}_*)$. Additionally, both $m(\mathbf{x}_*)$ and $s^2(\mathbf{x}_*)$ are influenced by Σ_ϵ: a generally larger noise variance (diagonals of Σ_ϵ) will dampen $\left[\mathbf{K}_f + \Sigma_\epsilon\right]^{-1}$, thereby causing $\mathbf{y}$ to influence the posterior mean less, and to increase the posterior variance. This matches the intuition that noisier data is less reliable for making predictions. In the limit $\sigma^2(\mathbf{x}_i) \to \infty$, the respective weight a_i in (1.10) goes to zero, nullifying the influence of y_i.

Noiseless ($\sigma^2(\mathbf{x}) = 0$) and noisy observations are treated in a unified manner by the GPs, and no modification is needed if any or all of the diagonal entries in Σ_ϵ are zero. Hence GPs offer an attractive apparatus that can be used both for interpolation (a common task in model calibration or more generally in "deterministic" experiments) and smoothing (arising due to noisy data or noisy estimates, e.g. from Monte Carlo methods). When an observation is exact, $\sigma^2(\mathbf{x}_i) = 0$, we have $y(\mathbf{x}_i) = f(\mathbf{x}_i)$ and there is no longer any uncertainty about $f(\mathbf{x}_i)$ at the given input $\mathbf{x}_i$. The fully noiseless case of $\sigma^2(\mathbf{x}) \equiv 0 \forall \mathbf{x}$ is common in practice where a deterministic (but often expensive to evaluate) function is available and an interpolating surface is to be modeled, such as computer experiments [81, 144], Bayesian optimization [123], and physical systems [171]. The GP surrogate then interpolates known f_i's, providing predictions and uncertainty quantification between observed $\mathbf{x}_i$. One may verify that if $(\mathbf{x}, y(\mathbf{x})) \in \mathcal{D}$, then the GP posterior moments are consistent with interpolation: $m(\mathbf{x}) = y(\mathbf{x})$ and $s(\mathbf{x}) = 0$.

In the noisy case, thanks to the assumption that $\epsilon(\mathbf{x}_i)$ are independent across i, one may straightforwardly allow for repeated inputs, i.e. $\mathbf{x}_i = \mathbf{x}_j$. Such replication corresponds to collecting information about the same functional response through repeated observations with independently generated noise, see Sect. 3.1.

Remark 1.4 All $\mathbf{x}_i \in \mathcal{X}$ are assumed known, i.e. non-random. From the probabilistic perspective, conditioning on $\mathcal{D}$ is thus the same as conditioning on $\mathbf{y}$, but emphasizes the input-output relationship $\mathbf{x} \mapsto y(\mathbf{x})$. For the case where inputs are uncertain, see Sect. 1.6.

The above derivation provides pointwise GP predictions. More generally, for any sequence of test points $\mathbf{x}_{1*}, \cdots \mathbf{x}_{N_**} \in \mathcal{X}$ the posterior vector satisfies $\mathbf{f}_* | \mathbf{y} \sim \mathcal{MVN}$:

$$[f(\mathbf{x}_{1*}), \cdots, f(\mathbf{x}_{N_**})]^\top | \mathcal{D} \sim \mathcal{MVN}\left(\mathbf{m}_*^\top, K(\mathbf{X}_*, \mathbf{X}_*)\right). \tag{1.11}$$

Applying multivariate Gaussian theory, the posterior mean function $\mathbf{m}_*$ is as before (replacing $K(\mathbf{x}_*, \mathbf{X})$ with the $N_* \times N$ matrix $K(\mathbf{X}_*, \mathbf{X})$), and the posterior covariance matrix is

$$\mathbf{K}_{f*} = K(\mathbf{X}_*, \mathbf{X}_*) - K(\mathbf{X}_*, \mathbf{X})\left[\mathbf{K}_y\right]^{-1} K(\mathbf{X}, \mathbf{X}_*). \tag{1.12}$$

If desired, the *posterior covariance function* can be obtained by considering two arbitrary $\mathbf{x}_*, \mathbf{x}_*'$, and $s(\mathbf{x}_*, \mathbf{x}_*') \stackrel{\triangle}{=} \text{cov}(f(\mathbf{x}_*), f(\mathbf{x}_*'))|\mathcal{D})$ is the off-diagonal of the matrix in Eq. (1.12). Once again, we invoke the Kolmogorov extension theorem to conclude that the *posterior process* $\{f_*(\mathbf{x})\}_{\mathbf{x} \in \mathcal{X}}$ is a Gaussian process, where by abusing notation we mean $f_*(\mathbf{x}) \stackrel{\triangle}{=} f(\mathbf{x})|\mathcal{D}$. Realizations of f_* yield posterior sample paths.

To illustrate the above concepts we present a simple example on $\mathbb{R}$. Here we use the quintessential squared exponential (SE) kernel, a common choice for GP models

$$k(x, x') \triangleq \exp\left(-\frac{1}{2}|x - x'|^2\right),\tag{1.13}$$

and the zero prior $\mu(x) = 0$. Note that this kernel is a function of the distance between its inputs $|x - x'|$, so that two observations that are spatially further apart produce a weaker covariance in their respective outputs. This leads to a smooth interpolation between data points. Our example supposes the synthetic dataset with $N = 5$ observations,

$$\mathcal{D} - \{(0.5, 1.0), (1.5, 0.5), (2.0, 0.75), (2.5, 0.25), (3.5, 0.0)\},$$

and constant observation variance $\sigma^2(\mathbf{x}) = \sigma_\epsilon$.

The left panel of Fig. 1.1 reflects the discussion in Sect. 1.2. In this setting it is the prior GP $f \sim \mathcal{GP}(0, k)$, representing initial beliefs about the output function in the absence of data. The sample path dynamics are governed by the covariance function; in this case, the smooth (infinitely differentiable) nature of the SE kernel is illustrated. As we observe in the left panel, this prior GP is described by its constant mean function $\mu(x) = 0$ and constant variance $\sigma^2(x) = 1$, reflecting a baseline level of uncertainty.

The shown sample paths are generated by sampling from the $\mathcal{MVN}$ for $\mathbf{y}(\mathbf{X}_*)$ based on the expression in (1.12). We pick a dense set of test points $\mathbf{x}_{1*}, \ldots, \mathbf{x}_{N_**}$ and then sample from the respective $\mathcal{MVN}$ with mean $(\mathbf{X}_*)^\top$ and covariance $\mathbf{K}_* \equiv K(\mathbf{X}_*, \mathbf{X}_*)$. In code this reduces to just calling a function like rmvnorm(mu=muStar, sigma=SigmaStar). Indeed, the GP framework neatly

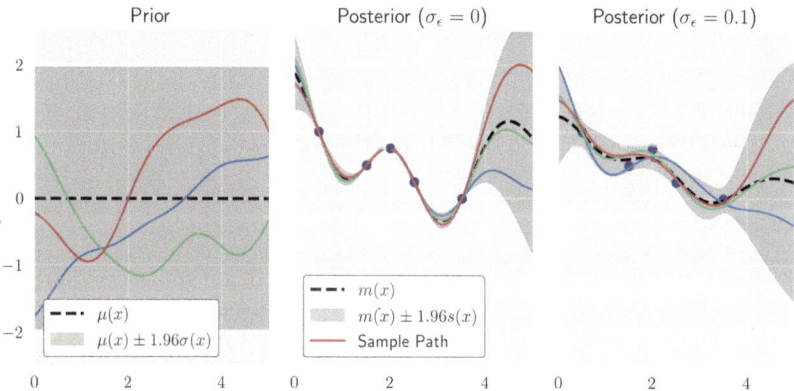

Fig. 1.1 Gaussian process regression using the SE kernel (1.13), visualized through three distinct scenarios. Left: Prior GP f. Middle: Posterior GP f_* with noiseless data ($\sigma_\epsilon = 0$, points). Right: Posterior GP f_* with noisy data ($\sigma_\epsilon = 0.1$, points). Each panel displays three sample paths, the mean, and the 95% uncertainty interval

fits with programming practices, whereby there is a natural equivalence between functions and vectors. Section 1.6 highlights some numerical and programming aspects.

Incorporating data (i.e. conditioning the prior on $\mathcal{D}$) and visualizing Eqs. (1.8) and (1.9), the middle and right panels illustrate the posterior GPs f_* when conditioned on five training points $\mathcal{D}$, for noiseless ($\sigma_\epsilon = 0$) and noisy ($\sigma_\epsilon = 0.1$) cases, respectively. Since observation variance is constant, $\Sigma_\epsilon = \sigma_\epsilon \mathbf{I}$. In the noiseless scenario, the posterior mean $m(\cdot)$ interpolates the observed data points exactly, whereas in the noisy scenario it captures the underlying trend while smoothing observations. Similarly, without noise $s^2(x_i) = 0$, while with noise the posterior variance is always strictly positive $s^2(x) > 0$ for all x.

The characteristic "sausage"-shaped uncertainty intervals seen in the posterior plots of Fig. 1.1 arise due to the SE kernel's properties, reflecting increased uncertainty in regions sparse in data. Importantly, as the predictions extend beyond the range of $\mathcal{D}$, the posterior reverts towards the prior $\mu(x) = 0$, a behavior that is particularly pronounced in the right panel. This reversion to the prior away from the observed data illustrates the core principle of GP regression—blending prior knowledge with observed data to make predictions while quantifying uncertainty. It is further reflected in the asymptotics $s^2(x_*) \to \sigma^2$, the prior variance, as x_* gets far from the training locations. Finally, note that noisy observations universally increase posterior uncertainty, so that the variance of f_* in the right panel is always strictly bigger than the variance in the middle panel.

GP Posteriors Via Bayes

An alternative derivation of the posterior f_* in (1.8)–(1.12) can be performed with Bayes' theorem directly. For simplicity, assume that all matrices involved are full rank and that the prior mean is zero, $\mu \equiv 0$. Denote the prior and conditional density functions of $\mathbf{f}$ and $\mathbf{f}$ given $\mathbf{y}$, as $p(\mathbf{f})$ and $p(\mathbf{f}|\mathbf{y})$ respectively, and similarly as $p(\mathbf{y})$, $p(\mathbf{y}|\mathbf{f})$ for $\mathbf{y}$ and $\mathbf{y}$ given $\mathbf{f}$. Note the following:

- Equation (1.5) shows $\mathbf{y}|\mathbf{f} \sim \mathcal{MVN}(\mathbf{f}, \Sigma_\epsilon)$.
- Equation (1.3) explicitly provides $p(\mathbf{f}) = (2\pi)^{-N/2} \frac{1}{\sqrt{|\mathbf{K}_f|}} \exp\left(-\frac{1}{2}\mathbf{X}^\top \mathbf{K}_f^{-1}\mathbf{X}\right)$.
- Integration of $p(\mathbf{y}|\mathbf{f}) \cdot p(\mathbf{f})$ shows the marginal density to be $\mathbf{y} \sim \mathcal{MVN}(\mathbf{0}, \mathbf{K}_y)$.

Therefore, a direct plug-in of Bayes' rule shows (after simplifying)

$$p(\mathbf{f}|\mathbf{y}) = \frac{p(\mathbf{y}|\mathbf{f}) \cdot p(\mathbf{f})}{p(\mathbf{y})} \propto \exp\left(-\frac{1}{2}(\mathbf{f} - \mathbf{m})^\top \left\{\mathbf{K}_f - \mathbf{K}_f \mathbf{K}_y^{-1} \mathbf{K}_f\right\}^{-1} (\mathbf{f} - \mathbf{m})\right),$$

$$(1.14)$$

(continued)

which implies that the posterior mean is $\mathbf{m}$ and the posterior covariance is $\mathbf{K}_f - \mathbf{K}_f \mathbf{K}_y^{-1} \mathbf{K}_f$. For a general $\mathbf{f}_*$, $p(\mathbf{f}_* | \mathbf{y})$ can be found through similar arguments.

Remark 1.5 The Gaussian assumption on ϵ is not required, but provides an easier derivation and closed form solutions. When Eq. (1.5) holds and the fdd of ϵ is given, the application of Bayes' rule in (1.14) can be repeated with more complicated expressions for $p(\mathbf{y}|\mathbf{f})$ and $p(\mathbf{y})$, see Chap. 3.

1.4 GP Likelihood and Hyperparameters

The GP predictive equations in the previous section provide a "recipe" to obtain a probabilistic model that lifts the data $\mathcal{D}$ into a function $m(\cdot)$ on $\mathcal{X}$ which is the conditional prediction, as well as a second function $s(\cdot)$ that provides uncertainty quantification. Observe how there is a formula (1.8) that unambiguously maps any training data to a prediction. This is similar to how a linear statistical model works (e.g. ordinary least squares multiple linear regression) where there are also linear algebra equations for the respective regression coefficients. One might decide that the surrogate construction is all done; but in reality everything hinges on the choice of the prior mean–covariance μ and k. The latter have to be provided by the modeler, same way as in multiple linear regression one must provide the basis functions.

How is this to be done? The answer is two-part. First, there are many different *kernel families*, i.e. specific parametric forms for k. For example, the SE kernel in (1.13) can be viewed as a special case of the parametric family of squared-exponential functions on $\mathbb{R}$:

$$k(x, x') \overset{\triangle}{=} \eta^2 \exp\left(-\frac{|x - x'|^2}{2\ell_{\text{len}}^2}\right). \tag{1.15}$$

Here, η^2 is the *outputscale* and ℓ_{len} the *characteristic lengthscale* or *lengthscale* for short. In Chap. 2 we provide many other families of kernels, and describe their properties and influence. Second, there is the *hyperparameter fitting* question: given a parametric kernel family, which element of that family is appropriate? E.g. what are appropriate values for η^2 and ℓ_{len}? Observe that in the earlier example in (1.13) we just took both to be 1; this was a convenient, unjustified choice.

Hyperparameter fitting brings us back to a classical statistical inference task. There are many options, foremost being maximum likelihood estimation that we address below. But we first remark that this is inherently a separate *auxiliary* problem. As indicated by the hyper-prefix, the choice of the particular k is often a meta-decision and one that does not drastically impact the outputted prediction.

For example, it can be completely appropriate in some settings to rely on *expert opinion*, meaning to have a hard-coded ("pre-tuned") kernel k that is used as-is for the prior covariance. Indeed, with experience in fitting many GPs and knowledge of how k impacts the regression, having a pre-tuned kernel can aid interpretability and (human) confidence in the surrogate. At the other end of the spectrum are fully Bayesian GPs which impose a further layer of Bayesian inference, starting with a hyper-prior on the kernel hyper-parameters and then eventually ending with hyper-parameter posteriors, capturing intrinsic uncertainty of the surrogate (i.e. the idea that many GP surrogates could do a good job on the given regression task).

To better understand the effect of varying hyperparameters, let us analyze the current setup, using k in Eq. (1.15). Including homoskedastic noise ($\Sigma_\epsilon = \sigma_\epsilon^2 \mathbf{I}$), the hyperparameters are $\boldsymbol{\theta} = (\eta^2, \ell_{\text{len}}, \sigma_\epsilon^2)$.

- The outputscale and noise variance parameters directly influence the *magnitude* of f and y: if $g(x)$ is a GP with (unit-outputscale) covariance kernel $\exp\left(-\left|x - x'\right|^2 / 2\ell_{\text{len}}^2\right)$ and $\epsilon_0(x) \sim \mathcal{N}(0, 1)$, one has the representation

$$f(x) = \eta g(x), \qquad \text{and} \qquad y(x) = \eta g(x) + \sigma_\epsilon \epsilon_0(x). \qquad (1.16)$$

- The outputscale is irrelevant to the *correlation function* of f:

$$c(x, x') \overset{\triangle}{=} \text{corr}(f(x), f(x')) = \frac{k(x, x')}{\sqrt{k(x, x)k(x', x')}} = \exp\left(-\frac{\left|x - x'\right|^2}{2\ell_{\text{len}}^2}\right). \qquad (1.17)$$

- The lengthscale ℓ_{len} is aptly named to "scale the length"; it is also sometimes referred to as a range parameter, especially in geostatistics. For example, contrasting $\ell_{\text{len}} = 1$ and $\ell_{\text{len}} = 100$ highlights whether a distance, say $|x - x'| = 10$, is considered "far" to the model. In the former, one finds $k(x, x') \approx 0$ and the latter $k(x, x') \approx 1$. The lengthscale controls correlation decay: it determines whether the GP has a slow decay of correlation in space, leading to visually smooth/slowly changing surfaces, or a fast decay that produces more wiggly fits.

1.4.1 Estimation and Likelihood

To obtain a good fit, we wish to optimize over $\boldsymbol{\theta}$. Since ℓ_{len} is embedded deep within coordinates of the covariance matrix, it cannot be analytically inferred, and learning it requires optimization of the likelihood. Maximum likelihoood estimation (MLE) is currently the de facto default for GP inference, its universality supported by a general preponderance in statistics of likelihood-based inference when distributional assumptions are made on the data, as with the MVN assumption driving GPs. MLE not only allows to reduce inference to optimization without any further criteria, but moreover provides nice properties of the resulting MLE estimator that helps with uncertainty quantification and enables use of metrics like the Bayesian information criterion (BIC). Nevertheless, we also mention that there are other non-MLE approaches to obtain the hyperparameters. Cross validation is common in many fields, to optimize $\boldsymbol{\theta}$ based on minimizing a "leave-one-out" error. This technique generalizes less well to higher dimensional hyperparameter spaces, whereas MLE tends to do well [79].

In geostatistical literature, a common technique is *variography* [33]. Variography is a hands-on, human expert-driven approach to GP fitting. The empirical semi-variogram bins the data by distance between inputs and computes sample covariances across those bins. Consider a sequence $0 = h_0 < h_1 < \ldots h_L$ which induces the distance stratification structure $I_\ell \doteq (h_{\ell-1}, h_\ell]$, $\ell = 1, \ldots, L$. For any distance h define the neighborhood $N(h) = \{(\mathbf{x}_i, \mathbf{x}_j) : \|\mathbf{x}_i - \mathbf{x}_j\| \in I(h)\}$ where $I(h)$ is the unique interval among I_ℓ's containing h (or $[h_L, \infty)$ if $h > h_L$). Set

$$\hat{\gamma}(h) \overset{\triangle}{=} \frac{1}{2|N(h)|} \sum_{(\mathbf{x}_i, \mathbf{x}_j) \in N(h)} (y_i - y_j)^2, \tag{1.18}$$

The resulting $\hat{\gamma}(h)$ is a step function in h. In variography, one visually matches $\hat{\gamma}(\cdot)$ with a power-exponential or Matérn function, e.g. $\gamma_\theta^{\text{exp}}(h) = \sigma^2 + \eta^2(1 - \exp(-(\frac{h}{R})^p))$, where σ^2 is the nugget, η^2 is the partial sill and R is the range. Observe that one can obtain a fitted semi-variogram from the GP surrogate, allowing a 1-to-1 comparison between the two [167].

Most common in the GP literature is maximum likelihood estimation (MLE). Recall that the observed data is the vector $\mathbf{y}$, which from Sect. 1.3 is known to have fdd $\mathbf{y} \sim \mathcal{MVN}(\mathbf{0}, \mathbf{K}_y)$, which leads to the likelihood $L(\boldsymbol{\theta}|\mathbf{y}) = p(\mathbf{y})$, where the hyperparameters $\boldsymbol{\theta}$ are embedded in $\mathbf{K}_y$:

$$L(\boldsymbol{\theta}|\mathbf{y}) = p(\mathbf{y}) = (2\pi)^{-N/2} \frac{1}{\sqrt{|\mathbf{K}_y|}} \exp\left(-\frac{1}{2}\mathbf{y}^\top \mathbf{K}_y^{-1}\mathbf{y}\right). \tag{1.19}$$

As is customary, taking logarithms yields the log-likelihood

$$\ell(\boldsymbol{\theta}|\mathbf{y}) = -\frac{N}{2}\log 2\pi - \frac{1}{2}\log|\mathbf{K}_y| - \frac{1}{2}\mathbf{y}^\top \mathbf{K}_y^{-1}\mathbf{y}, \tag{1.20}$$

viewed as a function of the hyperparameters. MLE looks for the minimizer $\boldsymbol{\theta}^*$ of ℓ, which is a nonlinear global optimization problem. Due to the presence of the determinant term $\log|\mathbf{K}_y|$ and the corresponding matrix inverse, this optimization is non-convex and often features a rough landscape for N large. To improve the numerics, we differentiate the log-likelihood in Eq. (1.20) with respect to a given hyperparameter θ_k yielding

$$\frac{\partial}{\partial \theta_k}\ell(\boldsymbol{\theta}|\mathbf{y}) = \frac{1}{2}\mathbf{y}^\top \mathbf{K}_y^{-1}\frac{\partial \mathbf{K}_y}{\partial \theta_k}\mathbf{K}_y^{-1}\mathbf{y} - \frac{1}{2}\text{trace}\left(\mathbf{K}_y^{-1}\frac{\partial \mathbf{K}_y}{\partial \theta_k}\right). \tag{1.21}$$

Further, a clever rescaling in (1.16) shows that $y(x) = \eta[g(x) + \gamma \epsilon_0(x)]$ where $\gamma = \sigma_\epsilon/\eta$. In this reparametrization, η can be thought of as an outputscale on the data generating process, and γ as a noise-to-signal ratio. In this case,

$$\mathbf{K}_y = \eta^2(\mathbf{C}_f + \gamma^2 \mathbf{I}_N) \quad \Rightarrow \quad \hat{\eta}^2_{MLE} = \frac{1}{N}\mathbf{y}^\top(\mathbf{C}_f + \gamma^2 \mathbf{I}_N)^{-1}\mathbf{y}, \qquad (1.22)$$

where $\mathbf{C}_f$ is the $N \times N$ correlation matrix of $\mathbf{f}$ and the implication follows from (1.21). Assuming zero prior mean, $\hat{\eta}^2_{MLE}$ is proportional to the mean residual sum of squares, providing the interpretation of outputscale as the typical amplitude of the response. The remaining MLEs of $\boldsymbol{\theta}$ do not have closed forms, but there are additional tricks along this line of reasoning [79]. Unless otherwise specified, the remainder of the text utilizes MLE estimates $\hat{\boldsymbol{\theta}}_{MLE}$.

The implementation of GP models entails inverting and computing the determinant of $\mathbf{K}_y$, which is of $O(N^3)$ complexity by default. Additionally, hyperparameter optimization features the determinant term $\log|\mathbf{K}_y|$ in Eq. (1.20), creating a non-convex optimization landscape. The presence of multiple modes can lead to overfitting; conversely the log-likelihood surface might feature flat ridges whereby a whole manifold of hyperparameters give similar likelihoods. Historically methods like NLOPT routines have been employed, but these are prone to getting trapped in local optima or requiring substantial computational resources. In contrast, modern implementations increasingly favor stochastic gradient descent (SGD) and its variants [145]. SGD, by its stochastic nature, can navigate the likelihood surface more effectively, often "convexifying" the underlying landscape, and thus finding better hyperparameters. SGD-based optimization in this context is typically "early-stopped" rather than run to convergence, acknowledging the inherent randomness in the SGD process and balancing the need for a robust solution against the risk of overfitting or excessive computation time.

Proper hyperparameter initialization before starting the optimizer is also key. A common solution is to scale inputs (dimension-wise) and outputs, so that the hyperparameter fitting is performed on e.g.,

$$x_{i,(j)} \mapsto \frac{x_{i,(j)} - \min(\mathbf{x}_{(j)})}{\max(\mathbf{x}_{(j)}) - \min(\mathbf{x}_{(j)})}, \qquad (1.23)$$

where $\mathbf{x}_{(j)}$ is the jth column of $\mathbf{X}$, or on a mean/std. dev transform. This can similarly be done for $\mathbf{y}$. Note that interpretation of the hyperparameters will generally change after rescaling, but some cases permit a reverse scaling. A common case is for parametric kernels of the form $k(x, x') = h((x - x')/\ell_{len})$ (like in Eq. (1.15)), where the original-scale $\ell_{len, orig}$ is obtained through $\ell_{len, orig, j} = a_j \cdot \ell_{len, j}$ if one used the standardization $x_{i,(j)} \mapsto (x_{i,(j)} - b_j)/a_j$.

Chapter 3 investigates further numerical aspects and improvements. This includes ways to reduce the $O(N^3)$ complexity, like sparse variational Gaussian processes [122], introducing sparsity in other ways, or using low-rank matrix approximations, and ways to address inversion of $\mathbf{K}_y$ which can have a high condition number when N is large or when many $\mathbf{x}_i$'s are spaced closely together when using a kernel like the SE kernel. In cases where the inferred σ_ϵ is very small, $\mathbf{K}_y$ can be near-singular leading to numerical instability. A common solution is regularizing the matrix inversion substep by adding a small *jitter* (say $\varsigma = 10^{-8}$) to the diagonal of $\mathbf{K}_y$, which ensures that $\mathbf{K}_y + \varsigma\mathbf{I}$ has a bounded condition number.

GPs can also efficiently handle *cross-validation*, which roughly speaking is temporarily removing data during model fitting and predicting on this removed portion to assess out-of-sample performance, repeating for the entire dataset. The (leave-one-out) LOO log predictive probability is defined as:

$$\ell(\boldsymbol{\theta}|\mathbf{y})_{\text{LOO}} \triangleq \sum_{i=1}^{N} \log p_{-i}(y_i|\boldsymbol{\theta}), \quad \log p_{-i}(y_i|\boldsymbol{\theta}) \triangleq -\frac{1}{2}\sigma_i^2 - \frac{(y_i - \mu_i)^2}{2\sigma_i^2} - \frac{1}{2}\log 2\pi,$$

where the notation $-i$ means removing the ith observation from $\mathcal{D}$. In other words, the LOO mean and variance μ_i and σ_i^2 are computed using Eqs. (1.8) and (1.9) trained on $\mathcal{D}\backslash\{y_i\}$ with subsequent inference at the left-out location $\mathbf{x}_i$. This can be done efficiently as:

$$\mu_i = y_i - \left[\mathbf{K}_y^{-1}\mathbf{y}\right]_i / [\mathbf{K}_y^{-1}]_{ii}, \qquad \sigma_i^2 = 1/[\mathbf{K}_y^{-1}]_{ii}, \tag{1.24}$$

so that $\mathbf{K}_y$ needs to be inverted just once, resulting in a negligible computational overhead, as compared to a naive implementation of refitting the GP N times. Note that the traditional LOO cross-validation squared error can be similarly found efficiently as $\frac{1}{N}\sum_{i=1}^{N}(y_i - \mu_i)^2$. The expressions in (1.24) can then be used for hyperparameter estimation, to minimize squared error, or to maximize the LOO likelihood.

Bayesian GPs

The so-called *fully Bayesian* GPs [60, 100] work with the hierarchical setup that augments the presented framework with a prior on the hyperparameters $\boldsymbol{\theta}$:

$$\boldsymbol{\theta} \sim p(\boldsymbol{\theta})$$

$$\mathbf{f}|\boldsymbol{\theta} \sim \mathcal{GP}(\mu, k_{\boldsymbol{\theta}})$$

$$\mathbf{y}|\mathbf{f} \sim \mathcal{N}(\mathbf{f}, \Sigma_\epsilon).$$

Integrating over the joint posterior gives

$$p(\mathbf{f}_*|\mathbf{y}) = \int\int p(\mathbf{f}_*|\mathbf{f}, \boldsymbol{\theta})p(\mathbf{f}|\boldsymbol{\theta}, \mathbf{y})p(\boldsymbol{\theta}|\mathbf{y})d\mathbf{f}d\boldsymbol{\theta} = \int p(\mathbf{f}_*|\mathbf{y}, \boldsymbol{\theta})p(\boldsymbol{\theta}|\mathbf{y})d\boldsymbol{\theta}$$

(continued)

where $p(\mathbf{f}_*|\mathbf{y}, \boldsymbol{\theta})$ is the standard GP posterior given $\boldsymbol{\theta}$. This implies that $p(\mathbf{f}_*|\mathbf{y})$ is a mixture of Gaussians, with the mixing according to $p(\boldsymbol{\theta}|\mathbf{y}) \propto p(\mathbf{y}|\boldsymbol{\theta})p(\boldsymbol{\theta})$. The latter is usually intractable and handled via Markov Chain Monte Carlo sampling. While Bayesian GPs are popular in some subfields of statistical learning, they are usually an overkill for financial applications due to the heavy computation required.

1.4.2 Examples and Discussion

In this section we utilize a synthetic univariate function to illustrate the GPR model's ability to recover the ground truth. Define

$$f_0(x) = -0.1x^2 + \sin(3x), \quad 0 \le x \le 6, \tag{1.25}$$

with noisy outputs $y(x) = f_0(x) + \epsilon$, $\epsilon \sim \mathcal{N}(0, 0.4^2)$. In order to compare aspects of GPs, we choose two designs $\mathcal{D}_N = \{(x_i, y_i)\}_{i=1}^N$, with $N = 20, 40$. The smaller $\mathcal{D}_{20}$ has uniformly spaced x_i's, with training design $\mathbf{x} = (1.0, 1.2, \ldots, 4.8)$, while $\mathcal{D}_{40} \supset \mathcal{D}_{20}$ contains additional inclusions:

- a finer partition over $[1, 2]$, including $x_i = 1.1, 1.3, 1.5, 1.7, 1.9$;
- replicated samples in $[3, 4)$, so that each of the five $x_i \in [3, 4)$ has $n_x = 3$ total samples, and
- an additional $n_x = 5$ samples for $x = 5$.

The primary goal is to recover $f_0(\cdot)$, based on the GP maximum a posteriori (MAP) estimate $m(\cdot)$. We apply the SE kernel (1.15) with zero prior mean $\mu(x) \equiv 0$. Hence three hyperparameters are estimated through MLE, $\ell_{\text{len}}, \eta^2$ and the observation noise σ^2. Figure 1.2 shows the resulting fits based on $\mathcal{D}_{20}$ and $\mathcal{D}_{40}$.

All throughout the interior interval $[1, 5]$, the fitted $m(\cdot)$ matches the general curvature of $f_0(\cdot)$ (dashed line) quite well. The models have excellent fits in the denser right region, especially for $x_* \in [3, 5]$ in the right panel with $N = 40$, where adding data has significantly improved the fit, especially at the local minimum near $3 < x_* < 4$ and around $x_* = 5$. However, in areas where no additional data was added, such as the local maximum near $2 < x_* < 3$, the fit remains unchanged—the added data is too far to significantly influence the fit in this region. Note that it is hard to judge the effect of finer additions (like in $[1, 2]$) versus repeated observations (like in $[3, 4)$), since the left panel already shows a good fit for $x_* \in [1, 2]$.

When data is sparse (or non-existent), prediction will be less accurate. This is particularly apparent near $x_* = 5$ for the left panel ($N = 20$). Notice that $m(x_*)$ does not extrapolate any data-driven pattern, and we should not expect it to! By nature of a "prior" mean, the GP is reverting to $\mu(x) = 0$ in the absence of

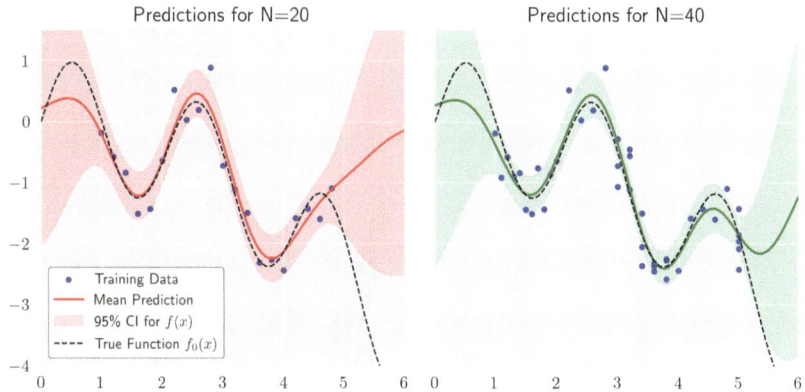

Fig. 1.2 GP fit illustrating the model's recovery of $f_0(x)$ and credible intervals

data informing the prediction, i.e., $m(x_*) \to 0$. This can be seen in Eq. (1.8) as $K(x_*, \mathbf{X}) \to \mathbf{0}$ since $|x_* - x_i| \to \infty$ for all $x \in \mathcal{D}$ (recall Eq. (1.15)). The right panel remedies this issue by including five observations at $x = 5$, extrapolating the observed decreasing pattern at $x = 5$. This decrease is projected to around $x = 5.5$, beyond which the predictive location is too far from $\mathcal{D}$.

The 95% confidence interval (CI) for $f(x_*)$ is computed as the $100(1 - \alpha)\%$ (pointwise) interval, given by $m(x_*) \pm z_{1-\alpha/2} \cdot s(x_*)$, where $z_{1-\alpha/2}$ is the standard normal quantile (in this case, $z_{0.975} = 1.96$). This calculation uses the posterior mean and standard deviation from (1.8) and (1.9). The intervals narrow in regions of increased data density (compare the panels for $1 \le x_* \le 2$ and $3 \le x_* \le 5$). The characteristic widening of the confidence interval near the training region boundaries ($x_* < 1$ and $x_* > 5$) reflects the scarcity of data, informing the modeler that prediction becomes highly unreliable outside of the range of $\mathcal{D}$, leading to an increase in $s^2(x_*)$ towards the prior variance $k(x_*, x_*) = \eta^2$. A prediction interval for $y(x_*)$ can be similarly derived as $m(x_*) \pm z_{1-\alpha/2}\sqrt{\sigma_\epsilon^2 + s^2(x_*)}$ using independence of $\epsilon(x_*)$.

Remark 1.6 Following [140], we do not distinguish between credible and confidence intervals. Parametric confidence intervals typically depend on distributional assumptions, analogous in our context to setting the prior mean and covariance kernel, along with a fixed $\mathcal{D}$. Data collection methods invariably introduce some deviations from these assumptions. Wang [165] discusses challenges related to the assumptions underlying GPs and the definition of confidence intervals, highlighting the impact of model misspecification.

Hyperparameter Sensitivity in GPR

Stepping back from MLE, we discuss the influence of hyperparameters, specifically lengthscales and outputscales, on the GP model's predictions.

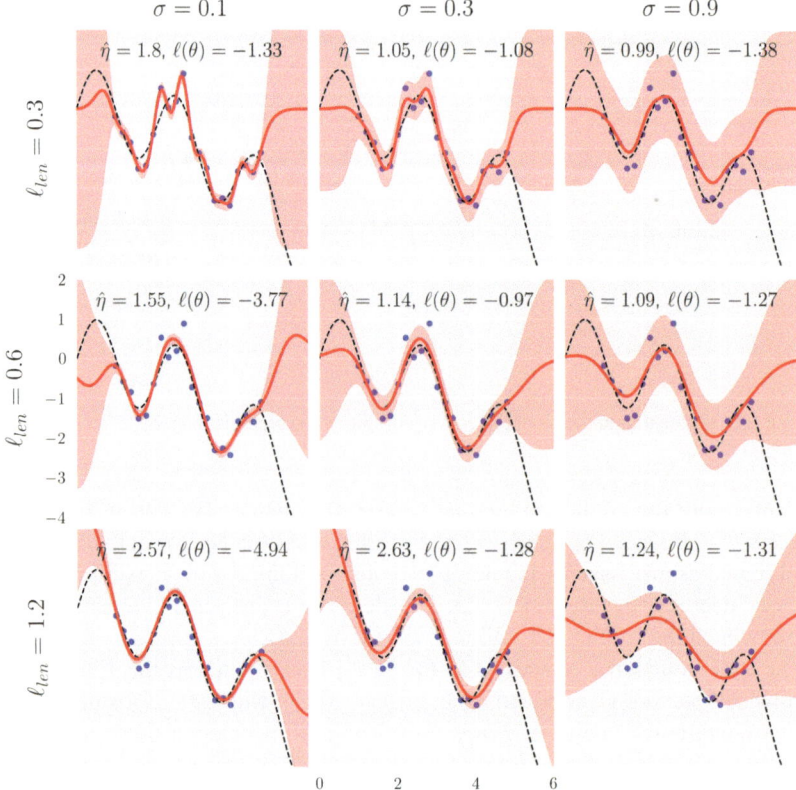

Fig. 1.3 Comparative effects of varying lengthscales and noises for the $\mathcal{D}_{20}$ dataset, with estimated outputscale $\hat{\eta}$ and corresponding likelihoods $\ell(\boldsymbol{\theta})$

In Fig. 1.3, hyperparameter configurations are compared for $\mathcal{D}_{20}$ in a three-by-three grid consisting of $\sigma = \{0.1, 0.3, 0.9\}$ and $\ell_{\text{len}} = \{0.3, 0.6, 1.2\}$, overlaying the value of $\hat{\eta}_{MLE}$ fitted through (1.22), and the corresponding likelihood value. The middle panel includes the (rounded-down) MLE values $\hat{\ell}_{\text{len}} = 0.6442$, $\hat{\sigma} = 0.3632$ (with $\hat{\eta} = 1.246$). For reference, the MLE estimate of $\boldsymbol{\theta}$ when fitted to $\mathcal{D}_{40}$ are $\hat{\ell}_{\text{len}} = 0.6764$, $\hat{\sigma} = 0.3758$ and $\hat{\eta} = 1.497$. Observe that while we have the ground truth $\sigma = 0.4$, there is no concept of a "true" ℓ_{len} and η for the fixed $f_0(x)$.

The general pattern is that as ℓ_{len} decreases, the model becomes more "local." This is especially true in the top left panel, where the model assumes little noise, so fluctuations in $y(x)$ are interpreted as coming from $f_0(x)$, a typical case of overfitting. A similar phenomenon, though less extravagant, is observed for $\sigma = 0.3$. In contrast, $\sigma = 0.9$ underfits or equivalently over-smoothes. Increasing σ across the columns generally dampens fluctuations in $m(\cdot)$, akin to thinking of the posterior mean as a string being pulled gently from both sides. The bottom right panel illustrates a case where both ℓ_{len} and σ are relatively large. This leads to an averaging of data, taking fluctuations as belonging to a long-run trend. Notice also that ℓ_{len} controls the rate of prior reversion: for $\ell_{len} = 0.3$, rapid convergence when extrapolating is clear where both $m(x)$ and $s(x)$ level off at the edges. This is not true as we move down the rows, noting a slower reversion to prior for $\ell_{len} = 1.2$.

Recalling that η controls the amplitude for $f_*(\cdot)$, one can visually estimate the relative size of $\hat{\eta}_{MLE}$ for a given panel based on comparing the minimum and maximum $m(\cdot)$ values. For example, the bottom right panel has a relatively small $\hat{\eta}_{MLE}$ since the posterior GP has narrower fluctuations. In contrast, the remaining panels in the bottom row lead to large $\hat{\eta}_{MLE}$ reflecting the wide fluctuations in $f_*(\cdot)$.

Estimating lengthscale and noise simultaneously corresponds to balancing between signal and noise. Long lengthscales are more common when noise is high, whereas short lengthscales offer the potential to explain away noise as quickly changing dynamics in the data. This tradeoff shows up as multimodal likelihood profile, see the upper-right to down-left diagonal in Fig. 1.3 which shows three reasonable fits. In this example, the middle panel (corresponding to the MLE estimates), appears the most visually appropriate, producing a close fit and reasonably sized confidence intervals, with no egregious over- or under-fitting.

1.4.3 Universal Kriging and Varying Prior Means

The linearity of the Gaussian distribution with respect to its mean enables organic incorporation of user-specified mean functions $\mu(\cdot)$ for the GP surrogate. Prior mean specification should be thought of as de-trending, with the same caveats. Namely, de-trending is not absolutely necessary but is a nice plus if one has a clear understanding of the trend.

Handling a $f \sim \mathcal{GP}(\mu, k)$ with known prior $\mathbb{E}[f(\mathbf{x})] = \mu(\mathbf{x})$ is straightforward, as one can just subtract off $\mu(\mathbf{x})$ and obtain a zero-mean GP. For example, the trend-corrected posterior mean is

$$m(\mathbf{x}_*) = K(\mathbf{x}_*, \mathbf{X}) [\mathbf{K} + \Sigma_\epsilon]^{-1} (\mathbf{y} - \boldsymbol{\mu}) + \mu(\mathbf{x}_*) \tag{1.26}$$

where $\boldsymbol{\mu} = [\mu(\mathbf{x}_1), \ldots, \mu(\mathbf{x}_N)]$ is the known prior mean at each input. In effect the posterior works by conditioning on the residuals between the observations $\mathbf{y}$ and their marginal means $\boldsymbol{\mu}$ on $\mathcal{D}$, and then adding the result back onto the marginal mean $\mu(\mathbf{x}_*)$ at the predictive location. The posterior variance remains as in (1.9) since there is no impact of the deterministic trend on uncertainty of f_*.

More generally, GPs allow to incorporate classical least-squares regression with known basis functions together with the "nonparametric" modeling. This can be understood as doing an additive model and then imposing a Gaussian Bayesian paradigm to learn the conditional distribution given the data. Consider L basis functions $h_1, \ldots, h_L$, amalgamated into the vector $\mathbf{h}(\mathbf{x}) = [h_1(\mathbf{x}), \ldots, h_L(\mathbf{x})]$. We wish to learn the respective coefficients $\boldsymbol{\beta} \in \mathbb{R}^L$, so that the overall trend would be $\sum_{\ell=1}^{L} \beta_\ell h_\ell(\cdot)$ or in matrix form $\mathbf{h}(\cdot)^\top \boldsymbol{\beta}$. The interpretation is that we have a global linear model captured by $\boldsymbol{\beta}$ and the residuals $\mathbf{y} - \mathbf{h}(\mathbf{x})^\top \boldsymbol{\beta}$ form a mean-zero GP. Joint optimization over $\boldsymbol{\beta}$ and over the GP yields the following *Universal Kriging* (UK) equations, [140, Section 2.6]

$$m(\mathbf{x}_*) = \mathbf{h}(\mathbf{x}_*)^\top \boldsymbol{\beta} + K(\mathbf{X}, \mathbf{x}_*)\mathbf{K}_y^{-1}(\mathbf{y} - \mathbf{H}\boldsymbol{\beta}) \tag{1.27}$$

$$s^2(\mathbf{x}_*) = k(\mathbf{x}_*, \mathbf{x}_*) + \mathbf{R}^\top (\mathbf{H}^\top \mathbf{K}_y^{-1}\mathbf{H})^{-1}\mathbf{R}, \tag{1.28}$$

where $\mathbf{H}$ is the $N \times L$ matrix collating the $\mathbf{h}$-row vectors and $\mathbf{R} = \mathbf{h}(\mathbf{x}_*) - \mathbf{H}^\top \mathbf{K}_y^{-1}\mathbf{H}$. Moreover the *generalized least squares estimator* for the coefficients $\boldsymbol{\beta}$ is

$$\hat{\boldsymbol{\beta}} = \left(\mathbf{H}^\top \mathbf{K}_y^{-1}\mathbf{H}\right)^{-1} \mathbf{H}^\top \mathbf{K}_y^{-1}\mathbf{y}, \tag{1.29}$$

matching the classical linear model estimator. From a Bayesian perspective (1.29) can be understood as putting a vague, infinite-variance prior $\boldsymbol{\beta} \sim \mathcal{N}(0, b\mathbf{I}_L)$ as $b \to \infty$, and (1.28) can be adjusted to incorporate an informative prior with $b < \infty$ above. Considering (1.27), we interpret $\mathbf{h}(\mathbf{x}_*)^\top \boldsymbol{\beta}$ as the mean of f prior to considering the observations; the overall prediction then combines this prior mean with a weighted average (weighted according to $\mathbf{a} = K(\mathbf{X}, \mathbf{x}_*)\mathbf{K}_y^{-1}$ as in (1.10)) of the mean-corrected observations $\mathbf{y} - \mathbf{H}\boldsymbol{\beta}$ to get the new conditional mean [167].

Constant Prior Mean

The canonical default choice in most software implementations of GPs is a constant (but non-zero) prior mean $\mu(x) = \beta_0$ which is then inferred with the other hyperparameters during the fitting stage. The respective estimate from (1.29), $\hat{\beta}_0 = \frac{1}{\mathbf{1}^\top \mathbf{K}_y^{-1}\mathbf{1}}\mathbf{1}^\top \mathbf{K}_y^{-1}\mathbf{y}$ corresponds to a kernel-weighted average response.

To illustrate the effect of $\mu(\cdot)$, we continue the example of Sect. 1.4.2, fitting four GPs based on $\mathcal{D}_{40}$ with varying polynomial mean functions, reflecting the feature that f_0 in (1.25) has a quadratic term. The resulting fits are shown in Fig. 1.4, and the hyperparameter MLEs are listed in Table 1.2. The in-sample fits are nearly indistinguishable and the reversion to the prior also occurs at nearly the same rate in all models. None of the models except for the quadratic mean are appropriate for extrapolating far beyond $x_* = 5$, and the f_* with quadratic $\mu(\cdot)$ also begins to diverge beyond $x_* > 7$. Despite extrapolation being generally difficult and dangerous, the GP remarkably obtains trend coefficients $\hat{\beta}_1 = 0.0353$ and $\hat{\beta}_2 = -0.0843$ (compare with the true $\beta_1 = 0$ and $\beta_2 = -0.1$ in (1.25)) even with a measly $N = 40$ data points contaminated by noise and a sinusoidal pattern. A prior mean that aligns closer with the truth also produces tighter confidence intervals, which happens monotonically in polynomial degree. Note further that the credible

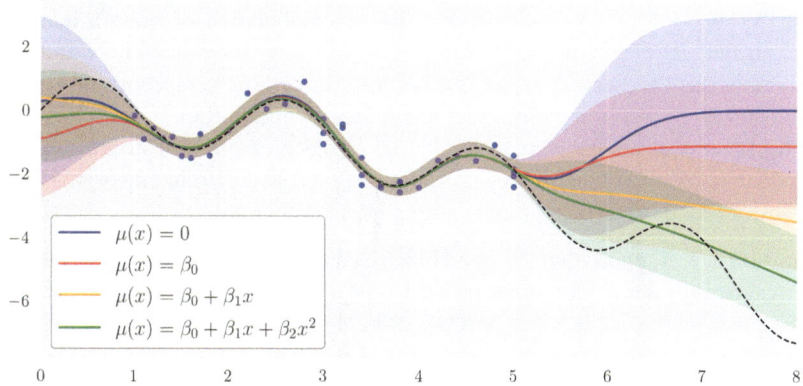

Fig. 1.4 Comparing $m(x)$ and 95% confidence intervals for varying prior mean function $\mu(\cdot)$ choices. All posterior GPs are based on the $\mathcal{D}_{40}$ dataset and the UK Eqs. (1.27)–(1.28). Ground truth (dotted black line) is $f_0(x) = -0.1x^2 + \sin(3x)$ in (1.25)

Table 1.2 Hyperparameter MLEs for GPs with varying prior means $\mu(\cdot)$ applied to $\mathcal{D}_{40}$

$\mu(x)$	$\hat{\sigma}$	$\hat{\ell}_{\text{len}}$	$\hat{\eta}$	$\hat{\beta}_0$	$\hat{\beta}_1$	$\hat{\beta}_2$
0	0.3758	0.6764	1.497	–	–	–
β_0	0.3780	0.5469	0.9681	-1.124	–	–
$\beta_0 + \beta_1 x$	0.3766	0.5002	0.7620	0.3413	-0.4809	–
$\beta_0 + \beta_1 x + \beta_2 x^2$	0.3764	0.4916	0.7469	-0.2813	0.0353	-0.0843

intervals for the model with $\mu(x) = 0$ are wider than for $\mu(x) = \beta_0$, mimicking the fact that the observations fluctuate further from their prior trend in that case. Thus, a better trend yields more confidence in the GPR outputs.

Looking at Table 1.2, note that the regression hypothesis can now be thought of as $y(x) = \mu(x) + f(x) + \epsilon(x)$ where $f(\cdot)$ is a mean-zero GP with outputscale η^2. Hence, a stable $\hat{\sigma}$ estimate is expected even when μ is changed, since $\epsilon(x)$ should not vary with x. On the other hand, the outputscale $\hat{\eta}$ is influenced by the choice of $\mu(\cdot)$ since some of the variation in $y(x)$ is now governed through $\mu(x)$. In particular, as the number of basis functions grows, $\hat{\eta}$ decreases, compare the respective rows in Table 1.2; this is the same phenomenon that decreases $s^2(\cdot)$ for more elaborate $\mu(\cdot)$. The estimated lengthscales also decrease across the rows, a result of longer-range trends being explained by the mean function and making the de-trended $f(\cdot)$ more wiggly.

1.5 GPs as Kernel Smoothing and Kernel Ridge Regression

GP regression inherently functions as a kernel smoother where the prediction $m(\mathbf{x}_*)$ at a new input $\mathbf{x}_*$ is a weighted sum of the observed data $\mathbf{y}$ with weights determined by a kernel function, see (1.8). The representation $m(\mathbf{x}_*) =$

$K(\mathbf{x}_*, \mathbf{X}) \left[\mathbf{K}_f + \Sigma_\epsilon \right]^{-1} \mathbf{y}$ in (1.10) as a linear combination of the y's quantifies the contributions of the respective training locations $\mathbf{x}_i$, modulated by k and noise characteristics. This section explores this kernel smoothing perspective on GPs and establishes its connection with Kernel Ridge Regression (KRR).

The functional analytic home for kernel based methods lies in the theory of reproducing kernel Hilbert spaces (RKHS). We present a distilled theory of RKHS and how it relates to GPs, beginning with a broad definition.

Definition 1.7 Suppose that $\mathcal{H}_k$ is a Hilbert space of real valued functions defined over a nonempty set $\mathcal{X}$ with inner product $\langle \cdot, \cdot \rangle_{\mathcal{H}_k}$ and $\| \cdot \|_{\mathcal{H}_k}$ as the norm induced by the inner product. Let $k : \mathcal{X} \times \mathcal{X} \to \mathbb{R}$ be a kernel. Then $\mathcal{H}_k$ is a *reproducing kernel Hilbert space* of $\mathcal{H}$ with *reproducing kernel* k if:

(a) For all $\mathbf{x}' \in \mathcal{X}$, the function $\mathbf{x} \mapsto k(\mathbf{x}, \mathbf{x}')$ is an element of $\mathcal{H}_k$;
(b) For every $\mathbf{x} \in \mathcal{X}$ and any function $g \in \mathcal{H}_k$, the evaluation $g(\mathbf{x})$ is given by the inner product $\langle g, k(\cdot, \mathbf{x}) \rangle_{\mathcal{H}_k}$.

In the GP modeling setting, it is common to desire dynamics according to a specified prior kernel. The Moore-Aronszajn Theorem ensures consistency with the RKHS framework:

Theorem 1.8 (Moore-Aronszajn Theorem) *For every positive definite kernel k, there exists a unique RKHS $\mathcal{H}_k$.*

The RKHS $\mathcal{H}_k$ thus provides a (unique) structured space capturing the influence of k. Briefly sketching the proof of this theorem highlights some connections with GP theory: for a given kernel k, let $\mathcal{H}_0$ be the span of kernel functions applied at points of $\mathcal{X}$:

$$\mathcal{H}_0 \overset{\triangle}{=} \mathrm{span}\{k(\cdot, \mathbf{x}) : \mathbf{x} \in \mathcal{X}\} = \left\{ \sum_{i=1}^{N} c_i k(\cdot, \mathbf{x}_i) : \exists\, N \in \mathbb{N}, \mathbf{x}_i \in \mathcal{X}, c_i \in \mathbb{R} \right\}.$$
(1.30)

One can verify that this defines an inner product (and induces a norm) $\langle g, h \rangle_{\mathcal{H}_0} = \sum_{i=1}^{N} \sum_{j=1}^{N} c_i d_j k(\mathbf{x}_i, \mathbf{x}_j)$ where the c_i and d_j are respective coefficients of g and h. The RKHS $\mathcal{H}_k$ is then the closure of $\mathcal{H}_0$ with respect to the norm $\| \cdot \|_{\mathcal{H}_0}$. One immediately sees that the posterior mean function of a GP with kernel k belongs to $\mathcal{H}_k$, since (cf. (1.10))

$$m(\mathbf{x}) = K(\mathbf{x}, \mathbf{X}) \mathbf{K}_y^{-1} \mathbf{y} = \sum_{i=1}^{N} a_i k(\mathbf{x}, \mathbf{x}_i) \in \mathcal{H}_0 \subseteq \mathcal{H}_k.$$

KRR is a form of regularized empirical risk minimization where the hypothesis space is an RKHS $\mathcal{H}_k$ associated with a kernel k. The objective in KRR is to find a function that minimizes the penalized residual sum of squares, balancing fidelity to the training data with the smoothness of the function (with respect to $\mathcal{H}_k$). Given the training $\mathbf{y}$, this is formalized as: $\hat{g}_{KRR} = \arg\min_{g \in \mathcal{H}_k} \frac{1}{N} \sum_{i=1}^{N} L(\mathbf{x}_i, y_i, g(\mathbf{x}_i)) + \lambda \|g\|_{\mathcal{H}_k}^2$, $\hat{g}_{KRR} = \arg\min_{g \in \mathcal{H}_k} \frac{1}{N} \sum_{i=1}^{N} L(\mathbf{x}_i, y_i, g(\mathbf{x}_i)) + \lambda \|g\|_{\mathcal{H}_k}^2$, where $L : \mathcal{X} \times \mathbb{R} \times \mathbb{R} \to \mathbb{R}^+$ is the loss function measuring the fit, and $\lambda > 0$ is the regularization factor penalizing the complexity of g. For the common squared loss case $L(\mathbf{x}, y, g) = |y - g(\mathbf{x})|^2$, this simplifies to

$$\hat{g}_{KRR} = \arg\min_{g \in \mathcal{H}_k} \frac{1}{N} \sum_{i=1}^{N} (y_i - g(\mathbf{x}_i))^2 + \lambda \|g\|_{\mathcal{H}_k}^2. \tag{1.31}$$

Despite $\mathcal{H}_k$ often being infinite-dimensional, the representer theorem [148] states that the solution of (1.31) is necessarily of the form $\hat{g}(\mathbf{x}) = \sum_{i=1}^{N} \alpha_i k(\mathbf{x}, \mathbf{x}_i)$, which leads to an explicit unique solution that is finite-dimensional:

$$\hat{g}_{KRR}(\cdot) = K(\cdot, \mathbf{X}) \left[\mathbf{K}_f + N\lambda \mathbf{I} \right]^{-1} \mathbf{y}. \tag{1.32}$$

This formulation of KRR bears a notable equivalence to the posterior mean function (1.8) in GPR. Specifically, for a given kernel k and a regularization parameter λ, the KRR solution $\hat{g}_{KRR}$ is equivalent to the posterior mean $m(\cdot)$ of a GP regression model $f \sim \mathcal{GP}(0, k)$ with the same kernel and noise variance $\sigma_\epsilon^2 = N\lambda$, $\epsilon_i \sim \mathcal{N}(0, N\lambda)$. This relationship offers a second viewpoint that the GP posterior mean $m \in \mathcal{H}_k$, since the KRR solution searches over functions in $\mathcal{H}_k$. Additionally, this characterization illustrates that σ_ϵ^2 (equivalently λ) affects the smoothness of the GPR predictions, with higher values leading to smoother posterior mean, interpreted as stronger penalization of the prediction's RKHS norm. Above, there is no mention of the GP distribution aside from its predictive (posterior) mean. Indeed, KRR only concerns itself with a single predictive function (which corresponds to the GP posterior mean), whereas the GP generates an entire stochastic process. This is where the equivalence breaks: generally speaking, GP sample paths do not belong to $\mathcal{H}_k$, but instead (almost surely) belong to a larger RKHS $\mathcal{H}_k^\theta$ called the *power RKHS* of k. In most cases, this RKHS contains less smooth functions than $\mathcal{H}_k$, see Sect. 2.1.

1.6 Closing Notes

GP Packages

There is a plethora of software packages implementing GP models [55]. A bare-bones implementation could be done in a couple of hours from first

(continued)

principles, since it only needs routine linear algebra operations and a nonlinear optimizer. For example, [79] walks through such steps using base R without any specialized tools. At the other end of the spectrum, advanced packages not only implement custom and accelerated hyperparameter optimization (for example relying on modern SGD suites) but also provide numerous variants of GPs, such as those discussed in Chaps. 2 and 3 below. First-generation packages were primarily written in R [78] or Matlab [139, 164]. With the advent of Python as the lingua franca of machine learning, the state-of-the-art has migrated to that environment, underpinned by the power of the Tensorflow and PyTorch optimization engines. Ultimately, there is no single "gold standard" GP package. We highlight the DiceKriging suite [144], laGP, hetGP [21] and kergp [46] in R, GPyTorch [65], GPy, GPflow and finally tinygp in Python and GP.jl in Julia.

On Numeric Stability

During hyperparameter optimization the $N \times N$ matrix $\mathbf{K}_y$ might end up with high condition numbers. To circumvent this, some libraries, such as GPyTorch [65], automatically (i.e. without user notification) incorporate jitter to ensure stable matrix inversion. On the other hand, packages like DiceKriging [144] assume no jitter, instead trusting the user that the expected σ_ϵ is large enough. Similarly, different implementations have varying defaults on when to terminate the MLE optimizer, whether to automatically scale inputs, etc. As a result, the respective fitted hyperparameters can vary, and reproducibility across packages can be elusive. The readers are thus cautioned to pay attention to optimizer nuances when using or comparing different GP softwares.

Further Reading

The main reference for Gaussian process theory is [3]. The "bible" for the modeling side of GP is the freely available monograph by Rasmussen and Williams [140]. A more modern reference is Gramacy [79] which emphasizes design of (computer) experiments and optimization.

Note that we assumed non-random $\mathbf{x}_i$ locations. Rasmussen and Williams [140, Section 9.5] discuss some ways to handle uncertain inputs. Chapter 2 discusses model selection, but it is worth noting that model mis-specification can manifest itself in various forms, for example with the hyperparameters being a nuisance parameter in confidence intervals for $f(\mathbf{x})|\mathcal{D}$. This has recently been thoroughly investigated in [165]. Reference [91] provides further commonalities and differences between GPs and KRR especially in minimizing the penalized residual sum of squares, and offers additional discussion of functional analysis for RKHS. We also highlight [59] who utilize KRR for stripping the discount curve, see Chap. 6.

This Chapter includes an online supplement consisting of a Python Jupyter notebook that reproduces the two case studies based on the synthetic univariate f_0 in (1.25).

Chapter 2
Covariance Kernels

2.1 First Examples and Smoothness

This chapter gives a more thorough examination of kernel functions $k(\mathbf{x}, \mathbf{x}')$ which are the primary driver of a GP model. The term kernel stems from its original use in the field of integral operators as studied by Hilbert and others, where the *kernel* is a function that gives rise to the integral operator $T_k g \overset{\Delta}{=} \int k(\cdot, \mathbf{x}) g(\mathbf{x}) d\mu(\mathbf{x}), g \in L_2(\mu)$ with respect to given measure μ. Kernels are the same positive definite functions used in other fields of machine learning, where their use is sometimes dubbed the "kernel trick," e.g. in support vector machines [148] or kernel PCA and clustering [150]. In this setting, the emphasis is that one can always write $k(\mathbf{x}, \mathbf{x}') = \langle \boldsymbol{\phi}(\mathbf{x}), \boldsymbol{\phi}(\mathbf{x}') \rangle_{\mathcal{H}}$ for some *feature mapping* $\boldsymbol{\phi}$ and inner product over a Hilbert space $\mathcal{H}$ [148, Proposition 2.11]. With this perspective, the underlying prediction (in our case, the posterior mean) given training inputs $\mathbf{x}_1, \ldots, \mathbf{x}_N$ is linear in the features:

$$m(\mathbf{x}) = \sum_{i=1}^{N} a_i k(\mathbf{x}, \mathbf{x}_i) = \sum_{i=1}^{N} a_i \langle \boldsymbol{\phi}(\mathbf{x}), \boldsymbol{\phi}(\mathbf{x}_i) \rangle_{\mathcal{H}}. \tag{2.1}$$

The elegance of this approach is that it allows to operate in a potentially high-dimensional feature space $\mathcal{H}$ without ever explicitly computing the features $\boldsymbol{\phi}(\mathbf{x})$; instead, the inner product $\langle \boldsymbol{\phi}(\mathbf{x}), \boldsymbol{\phi}(\mathbf{x}_i) \rangle_{\mathcal{H}}$ is evaluated implicitly through $k(\mathbf{x}, \mathbf{x}')$. Since many cases are high- or even infinite-dimensional (e.g. SE kernel [140, Section 4.2.1]), the kernel trick generally reduces computational complexity while allowing to learn non-linear relationships in the input space.

The feature space representation provides rationale for $k(\mathbf{x}, \mathbf{x}')$ to be a *similarity measure* due to the respective correspondence with an inner product in a Hilbert space. Recalling that for $f \sim \mathcal{GP}(0, k)$, $k(\mathbf{x}, \mathbf{x}') = \text{cov}(f(\mathbf{x}), f(\mathbf{x}'))$, the choice of

© The Author(s), under exclusive license to Springer Nature Switzerland AG 2025
M. Ludkovski, J. Risk, *Gaussian Process Models for Quantitative Finance*,
SpringerBriefs in Quantitative Finance,
https://doi.org/10.1007/978-3-031-80874-6_2

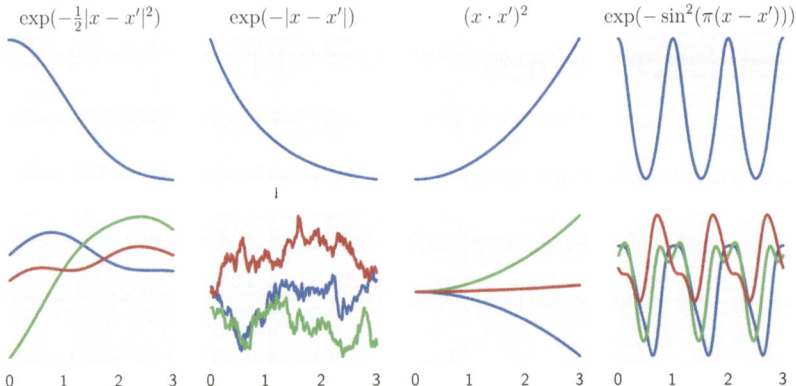

Fig. 2.1 Four different GP kernels on $\mathbb{R}$. Top row: prior kernel function $k(x, x')$ versus $x - x' \in [0, 3]$ except for $x \cdot x'$ which plots $k(x, 1)$ over $x \in [0, 3]$. Bottom row: three prior sample paths of $f(x)$, $x \in [0, 3]$ for each respective kernel

k manifests different properties in the prior (and consequently, posterior) GP sample paths. Figure 2.1 compares prior kernel functions and associated prior sample paths on $\mathbb{R}$ for four varying flavors of similarity. The SE kernel (1.15) (first column) measures similarity in terms of an exponential squared distance (further x and x' have more dissimilar outputs). The second column has a similar monotonic decay via $\exp(-|x - x'|)$, and while its sample paths are continuous, they are nowhere differentiable, unlike the SE which yields infinitely differentiable sample paths. The third column is the quadratic kernel $x \cdot x'$, which generates sample paths of the form $f(x) = ax^2$ for some $a \in \mathbb{R}$, providing a quadratic global structure. For the last column, its kernel is a function of $x - x'$ like the first two, but is not monotonically decreasing. Instead $k(x, x + h) = 1$ for $h = 1, 2, \ldots$. This repeating pattern produces sample paths that are periodic, $f(x) = f(x + 1)$.

To begin comparing the properties of different kernels, we quantify the behavior of GP paths. One aspect is the overall smoothness, summarized in terms of differentiability. Denote $C^m(\mathbb{R}^d)$ as the space of m-times differentiable functions over $\mathbb{R}^d$ whose derivatives are continuous. The space $C^0(\mathbb{R}^d)$ is the space of continuous, but not necessarily differentiable functions. The following well-known result relates regularity of the sample paths of $f \sim \mathcal{GP}(0, k)$ to properties of its kernel k [3, Theorem 2.2.2]. Here, we use the shorthand ∂_{x_j} and $\partial^2_{x_j, x_k}$ for the operators $\partial / \partial x_j$ (resp. $\partial^2 / (\partial x_j \partial x_k)$).

Proposition 2.1 *Suppose* $f \sim \mathcal{GP}(0, k)$. *If the derivative* $\partial^2_{x_j, x'_j} k(\mathbf{x}, \mathbf{x}')$ *exists and is finite at* $(\mathbf{x}, \mathbf{x}) \in \mathbb{R}^{2d}$, *then the limit* $\partial_{x_j} f(\mathbf{x}) \overset{\triangle}{=} \lim_{h \to 0} (f(\mathbf{x} + h\boldsymbol{\delta}_j) - f(\mathbf{x})) / h$ *exists in mean-square (m.s.). Conversely if the above limit exists for each* $\mathbf{x} \in \mathbb{R}^d$, *then* $\partial_{x_j} f(\cdot)$ *is the m.s. derivative of* $f(\cdot)$ *with kernel* $\partial^2_{x_j, x'_j} k(\mathbf{x}, \mathbf{x}')$.

Here, $\boldsymbol{\delta}_j \in \mathbb{R}^d$ is the vector with 1 at the jth coordinate and 0 otherwise. Since differentiation is a linear operator, it is further true that $\partial_{x_j} f \sim \mathcal{GP}(0, \partial^2_{x_j, x'_j} k(\mathbf{x}, \mathbf{x}'))$, and one may also infer the joint distribution of $[f, \partial_{x_j} f]^\top$ (see [140, Section 9.4]), allowing for analysis involving the derivative of a GP, see Sect. 4.3.1.

The notion of sample paths' differentiability is distinct from that of the GP posterior mean. For example, Fig. 2.1 showcased $f \sim \mathcal{GP}(0, k)$ with $k(x, x') = \exp(-|x - x'|)$ to produce non-differentiable $f \in C^0(\mathbb{R})$. However, recalling Equation (1.8), the respective posterior mean is of the form $m(x) = \sum_{i=1}^N a_i k(x, x_i) = \sum_{i=1}^N a_i \exp(-|x - x_i|)$, which is differentiable almost everywhere. More precisely, recall Sobolev spaces $H^m(X)$ of functions defined on $X \subseteq \mathbb{R}^d$ which have weak derivatives up to order m (see Appendix A.3.1 for a mathematical background). Given a kernel k, the elements of $\mathcal{H}_k$ (like $m(\cdot)$) are typically "smoother" than draws from $\mathcal{GP}(0, k)$ [162], in the sense that if $\mathcal{H}_k = H^m(X)$, then the GP sample paths almost surely belong to an RKHS that is (norm equivalent to) less smooth $H^{m-d/2}(X)$.

Another notion of smoothness can be defined through Hölder continuity. Suppose $f \sim \mathcal{GP}(0, k)$ over $X \subseteq \mathbb{R}^d$ has stationary increments. If the asymptotic relationship $\mathbb{E}[|f(\mathbf{x}+\mathbf{h}) - f(\mathbf{x})|^2] \sim c\|\mathbf{h}\|^\alpha$ holds for $c > 0$ and some $\alpha \in (0, 2)$ as $\|\mathbf{h}\| \to 0$, then f has a modification with locally ϕ-Hölder continuous trajectories for any $\phi \in (0, \alpha/2)$ (e.g. [15, Proposition 2.1]). The parameter α can be thought of as a *roughness index*, with smaller values indicating rougher realizations (recall that standard Brownian motion is locally ϕ-Hölder continuous for $\phi \in (0, 1/2)$, corresponding to $\alpha = 1$). The quantity α is sometimes linked to the *fractal dimension* $D = d + 1 - \frac{\alpha}{2}$ [72], which coincides with the Hausdorff dimension of the graph of f, $\mathrm{Gr} f \overset{\triangle}{=} \{(\mathbf{x}, f(\mathbf{x})) : \mathbf{x} \in X\} \subset \mathbb{R}^{d+1}$ [3, Theorem 8.4.1].

2.2 Classes and Properties of Kernels

The following classes of kernels $k(\mathbf{x}, \mathbf{x}')$ will be mentioned repeatedly:

- A kernel is *stationary* if it is a function of only $\mathbf{x} - \mathbf{x}'$, thus being invariant to translations in the input space. Such kernels may be written as $k(\mathbf{h})$ where $\mathbf{h} \overset{\triangle}{=} \mathbf{x} - \mathbf{x}'$ is the difference between inputs.
- An *isotropic* kernel is one that only depends on the distance $r = \|h\| = \|\mathbf{x} - \mathbf{x}'\|$. In this case, we may write $k(r)$ if isotropy is obvious from context. Note that isotropic kernels are stationary, but the converse is only true over $\mathbb{R}$ (the one-dimensional case).
- A kernel is *separable* if it can be written as a product of two kernels with no shared dimensions. For example, on $\mathbb{R}^2$ with $\mathbf{x} = (x_1, x_2)$, $k(\mathbf{x}, \mathbf{x}') = k_1(x_1, x'_1)k_2(x_2, x'_2)$ is separable if k_1 and k_2 are kernels on $\mathbb{R}$.

One key attraction of GP regression is that its methodology can be made agnostic to the problem dimension by utilizing separability, organically handling different $\mathcal{X}$. To wit, the standard way to construct kernels on $\mathbb{R}^d$ is via products: for $\mathbf{x} = (x_1, \ldots, x_d)$ we take

$$k(\mathbf{x}, \mathbf{x}') = \prod_{j=1}^{d} k_j(x_j, x_j'),$$

where k_j are univariate kernels on $\mathbb{R}$, sometimes called *one-dimensional base kernels*, with hyperparameters $\boldsymbol{\theta}_j$. In principle, one can arbitrarily mix different kernel families together, but in practice it is most common to use the same family for all coordinates, so that $k(\mathbf{x}, \mathbf{x}'; \boldsymbol{\theta}) = \eta^2 \prod_{j=1}^{d} k_{\text{base}}(x_j, x_j'; \theta_j)$ where $\boldsymbol{\theta} = (\theta_1, \ldots, \theta_d)$ is the concatenated vector of the GP hyperparameters over each dimension. Note that the product structure implies there is a single scale η^2 up-front, along with d different θ_j hyperparameters. When k_{base} is stationary and normalized so $k_{\text{base}}(x, x) = 1$, it follows that $\text{var}(f(\mathbf{x})) = \eta^2$ is the process variance. Isotropic kernels, in contrast, treat all coordinates symmetrically (appropriate when there is, say, spatial symmetry), $\theta_j \equiv \theta$.

2.2.1 Stationary Kernels

An important result in the theory of stationary kernels links them to spectral densities via Bochner's theorem. We state the version from [140].

Theorem 2.2 (Bochner's Theorem) *A complex valued function k on $\mathbb{R}^d$ is the covariance function of a weakly stationary mean-square continuous complex-valued random process $\mathbb{R}^d$ if and only if it can be represented as*

$$k(\mathbf{h}) = \int_{\mathbb{R}^d} e^{2\pi i \mathbf{u} \mathbf{h}} d\mu(\mathbf{u}), \tag{2.2}$$

where μ is a positive finite measure.

If μ in Eq. (2.2) has a density $S(\cdot)$, then S is known as the *spectral density* or *power spectrum* of k. In this case, k and S are Fourier duals of one another (known as the Wiener-Khintchine theorem),

$$k(\mathbf{h}) = \int S(\mathbf{u}) e^{2\pi i \mathbf{u} \mathbf{h}} d\mathbf{u}, \qquad S(\mathbf{u}) = \int k(\mathbf{h}) e^{-2\pi i \mathbf{u} \mathbf{h}} d\mathbf{h}. \tag{2.3}$$

Hence, one can construct kernels by taking a probability density function $S(\cdot)$, and applying (2.3). Note that stationary kernels satisfy $k(\mathbf{x}, \mathbf{x}) = k(\mathbf{0}) = \eta^2$, meaning that $k(\mathbf{h})/k(\mathbf{0})$ is a correlation function.

A common hyperparameter in stationary kernels is the lengthscale which appears in the form of $(\mathbf{x} - \mathbf{x}')/\ell_{len}$, reflecting the scaling of the inputs relative to the canonical univariate $[0, 1]$ domain. In order to translate lengthscales between different families, we recall a standardization based on the number of *upcrossings* of an $\mathbb{R}$-valued GP sample path. In short, more upcrossings lead to more fluctuations and hence a more apparent oscillation or wiggliness. Recall that an upcrossing of level u occurs at x_0 if $f(x_0) = u$ and $f'(x_0) > 0$. Let N_u denote the random variable counting the number of upcrossings of a level u over $x \in [0, 1]$ of $f \sim \mathcal{GP}(0, k)$ where k is stationary. If k is twice differentiable, the expected number of upcrossings is given by [140]

$$\mathbb{E}[N_u] = \frac{1}{2\pi} \sqrt{\frac{-k''(0)}{k(0)}} \exp\left(-\frac{u^2}{2k(0)}\right). \tag{2.4}$$

If $k''(0)$ does not exist (so that the process is not mean square differentiable) then if the respective GP f hits level u at x_0, it will almost surely have an infinite number of upcrossings of u in the arbitrarily small interval $(x_0, x_0 + \delta)$. Taking $u = 0$ and fixing the process variance $k(0) = \eta^2$, we see that the number of upcrossings is driven by $\sqrt{-k''(0)}$, which in turn is controlled by the kernel hyperparameters. This allows for an interpretation of the lengthscale in terms of $\mathbb{E}[N_u]$. For example, a single upcrossing is expected when $k''(0) = -4\pi^2\eta$, so we can use that metric to recover "equivalent" lengthscales as the kernel family is changed.

Long-range dependence is connected with the asymptotic behavior of the stationary kernel $k(\mathbf{h})$. In particular, if

$$k(\mathbf{h}) \sim \|\mathbf{h}\|^{-\beta} \qquad \text{as } \|\mathbf{h}\| \to \infty,$$

holds for some $\beta \in (0, 1)$, the GP is said to have *long memory* with Hurst coefficient $H = 1 - \frac{\beta}{2}$ [16, 72]. Note that this implies $\int_0^\infty |k(\mathbf{h})| \, d\mathbf{h} = \infty$.

We proceed to present several popular families of stationary kernels, summarizing their main features.

SE Kernels We have already utilized this kernel over $\mathbb{R}$ in Chap. 1. The squared exponential (SE) kernel goes by many alternative names, including the *radial basis function (RBF)* and *Gaussian* kernel (resembling the standard Gaussian pdf). The isotropic SE kernel over $\mathbb{R}^d$ is defined as

$$k(\mathbf{x}, \mathbf{x}') = \prod_{j=1}^{d} \exp\left(-\frac{1}{2\ell_{len}^2}|x_j - x_j'|^2\right)$$

$$= \exp\left(-\frac{1}{2\ell_{len}^2}\sum_{j=1}^{d}|x_j - x_j'|^2\right) = \exp\left(-\frac{r^2}{2\ell_{len}^2}\right) \tag{2.5}$$

where $r^2 = \|\mathbf{x} - \mathbf{x}'\|^2$. Hence the multi-dimensional SE kernel can be understood as a *univariate* SE kernel $k(r) = \exp(-r^2/2\theta^2)$ that works with the Euclidean norm $\|\mathbf{x} - \mathbf{x}'\|$ in $\mathbb{R}^d$, giving the name of radial bases. It has spectral density $S(\mathbf{u}) = (\sqrt{2\pi}\ell_{\text{len}})^d \exp(-2\pi^2\ell_{\text{len}}^2\mathbf{u}^2)$.

The (anisotropic) *generalized SE kernel* is a product of SE kernels with varying lengthscales per dimension

$$k(\mathbf{h}) = \exp(-\frac{1}{2}\mathbf{h}^\top\mathbf{A}^{-1}\mathbf{h}), \tag{2.6}$$

where $\mathbf{A}$ is a positive definite matrix. The separable case is $\mathbf{A} = \text{diag}(\ell_1^2, \ldots, \ell_d^2)$, and the isotropic case is $\mathbf{A} = \ell_{\text{len}}\mathbf{I}_d$. A GP *Single Index Model* uses (2.6) but constrains $\mathbf{A}$ to be of rank 1. The anisotropic separable SE is also known as the *automatic relevance determination (ARD)* kernel, usually re-written as

$$k_{ARD}(\mathbf{x}, \mathbf{x}'; \ell_1, \ldots, \ell_d) \overset{\triangle}{=} \eta^2 \exp\left(-\frac{1}{2}\tilde{r}(\mathbf{x}, \mathbf{x}')^2\right),$$

$$\tilde{r}(\mathbf{x}, \mathbf{x}')^2 = \|\mathbf{x} - \mathbf{x}'\|_\mathbf{A}^2 = \sum_{j=1}^{d}(x_j - x_j')^2/\ell_j^2. \tag{2.7}$$

The ARD name reflects the aspiration that the data can inform which coordinates have longer lengthscales (and hence are less relevant in terms of impact on the output) and which have shorter lengthscales and are more relevant. Through estimation, irrelevant dimensions have $\ell_j^2 \to \infty$, yielding $\exp(-\frac{1}{2\ell_j^2}|x_j - x_j'|^2) \approx 1$ to have negligible effect on the product kernel k_{ARD}. The fact that the ARD kernel can be viewed as both the scalar SE applied to the dimension-weighted Euclidean distance $\tilde{r}(\mathbf{x}, \mathbf{x}')^2$ and as a product of univariate SE kernels with distinct lengthscales is unique to this family and does not hold otherwise.

SE Parameterization
Generally speaking, there is no agreed-upon parametrization for the SE kernel lengthscale. For example, one may see $\exp(-r^2/\theta)$, $\exp(-\theta r^2)$, $\exp(-r^2/(2\theta))$, $\exp(-r^2/(2\ell^2))$, and $\exp(-(r/\ell)^2)$, with all permutations appearing in the literature. Thus, there is no standard on whether the hyperparameter is in the numerator or denominator, whether it is squared or not, and whether there is a normalizing factor of two, so when reading one must be careful to confirm what version is being assumed. In this book we use the probabilistic perspective with the "Gaussian" parameterization $\exp(-r^2/(2\ell^2))$.

Matérn Class A popular class of stationary kernels is the Matérn class, tracing back to [126] (see also [137, 156]). In the general univariate case, these kernels are defined by

$$k(r; \nu, \ell_{\text{len}}) = \frac{2^{1-\nu}}{\Gamma(\nu)} \left(r \sqrt{\frac{2\nu}{\ell_{len}}} \right)^{\nu} K_{\nu} \left(r \sqrt{\frac{2\nu}{\ell_{len}}} \right), \tag{2.8}$$

where K_{ν} is the modified Bessel function of the second kind. The respective spectral density of a one-dimensional Matérn-ν GP is $S_{\nu}(u) = 2\sqrt{\pi} \frac{\Gamma(\nu+1/2)}{\Gamma(\nu)} (2\nu \ell_{\text{len}}^{-2})^{\nu}$ $(2\nu \ell_{\text{len}}^{-2} + 4\pi^2 u^2)^{-\nu-1/2}$ [83]. In the limit $\nu \to \infty$, we have $S_{\nu}(u) \to \exp(-2\pi^2 \ell_{\text{len}}^{-2} u^2)$ which shows that the limiting kernel is the SE. For the multivariate version, one can take a product of univariate Matérn's (with distinct $\nu_1, \ldots, \nu_d$ if desired), or a Matérn ARD kernel, $k(\mathbf{x}, \mathbf{x}') = k_{\nu}(\|\mathbf{x} - \mathbf{x}'\|_{\mathbf{A}})$ [154].

The parameter ν in the Matérn-ν corresponds to smoothness: a GP on $\mathbb{R}$ with a Matérn-ν kernel is $\lceil \nu \rceil - 1$ times differentiable in the mean-square sense [140]. Hence the SE as a limiting case is infinitely differentiable. The sample paths of $f \sim \mathcal{GP}(0, k_{\nu})$ belong to the Sobolev space H^{ν} and the RKHS of k_{ν} is $H^{\nu+d/2}$ (recall that the sample path RKHS is less smooth than $\mathcal{H}_k$). As we will see in Sect. 2.5, the Matérn class of kernels produce "convergence" of their posterior GP as $N \to \infty$, under the assumption that the underlying $f_0 \in H^{\beta}$ for some β. In that case, the optimal rate is achieved when the prior kernel has $\nu = \beta$, meaning the GP sample path smoothness coincides with that of the regression function being modelled.

In practice, ν is fixed. This is partly due to the form (2.8), which bears a more tractable form for special cases of ν (particularly, if $\nu = p + 1/2$ where $p = 0, 1, 2, \ldots$). Common are

$$\nu = 1/2: \quad k_{M12}(r; \ell_{\text{len}}) \triangleq \exp\left(-\frac{1}{\ell_{\text{len}}} r \right), \tag{2.9}$$

$$\nu = 3/2: \quad k_{M32}(r; \ell_{\text{len}}) \triangleq \left(1 + \frac{\sqrt{3}r}{\ell_{\text{len}}} \right) \exp\left(-\frac{\sqrt{3}}{\ell_{\text{len}}} r \right), \tag{2.10}$$

$$\nu = 5/2: \quad k_{M52}(r; \ell_{\text{len}}) \triangleq \left(1 + \frac{\sqrt{5}r}{\ell_{\text{len}}} + \frac{5r^2}{3\ell_{\text{len}}^2} \right) \exp\left(-\frac{\sqrt{5}}{\ell_{\text{len}}} r \right), \tag{2.11}$$

resulting in GPs on $\mathbb{R}$ with sample paths that are non-differentiable ($\nu = 1/2$), one-time ($\nu = 3/2$), and two-times differentiable ($\nu = 5/2$). The $\nu = 1/2$ case recovers the Ornstein-Uhlenbeck process [19], which is mean reverting and continuous but non-differentiable. The mean-reversion of k_{M12} localizes dependence and has been advocated [88] for capturing small-scale effects. The lengthscale ℓ_{len} controls the rate of mean-reversion (lower values revert more quickly).

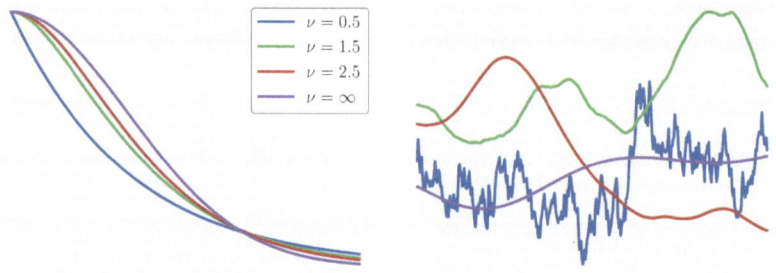

Fig. 2.2 Varying ν for Matérn family of kernels. Left: prior kernel $k_\nu(r)$ versus $r \in [0, 3]$ Right: Corresponding prior sample paths of $f(x)$, $x \in [0, 3]$

Figure 2.2 shows $k_\nu(r)$ and corresponding sample paths of $f \sim \mathcal{GP}(0, k_\nu)$ for $\nu \in \{0.5, 1.5, 2.5, \infty\}$ all with $\ell_{\text{len}} = 1$; recall that the ∞ case identically produces the SE kernel. A monotonic change in smoothness is observed in comparing the sample paths. Note that the kernel functions themselves all look similarly, highlighting that subtle differences in the kernel shape may produce drastically different behavior in the underlying process.

Rational Quadratic   The rational quadratic (RQ) kernel is defined as

$$k_{RQ}(r; \alpha, \ell_{\text{len}}) \triangleq \left(1 + \frac{r^2}{2\alpha\ell_{\text{len}}^2}\right)^{-\alpha}. \tag{2.12}$$

One way to interpret the RQ kernel is as a marginalized version of the SE kernel with a Gamma prior on the inverse-squared lengthscale $1/\ell_{\text{SE}}^2$ (with shape α and rate ℓ_{RQ}^2). Reparameterizing $\tau = 1/\ell_{\text{SE}}^2$,

$$\int_0^\infty \exp\left(-\frac{\tau r^2}{2}\right) \cdot p_{\text{Gamma}}(\tau; \alpha, \ell_{RQ}^2) d\tau \; \propto \; \left(1 + \frac{r^2}{2\alpha\ell_{RQ}^2}\right)^{-\alpha}.$$

Lower α places additional uncertainty on the lengthscale, resulting in more "wiggly" sample paths. A special case is $\alpha = 1$ in (2.12), resulting in the *Cauchy kernel*, named after the Cauchy pdf of the same form

$$k_{Chy}(r; \ell_{\text{len}}) \triangleq \left(1 + \frac{r^2}{2\ell_{\text{len}}^2}\right)^{-1}. \tag{2.13}$$

This "heavy-tailed" kernel models long-range dependence, meaning that correlations decay not exponentially but polynomially, leading to a long range influence between inputs [88]. The left column of Fig. 2.3 visualizes $k_{RQ}(r)$ and its sample paths for $\alpha \in \{0.2, 1, 5\}$, each with $\ell_{\text{len}} = 1$. Note that the $\alpha = \infty$ case is the SE kernel (see Figs. 2.1 and 2.2). As expected, more erratic patterns are seen in sample paths for lower α.

Generalized Cauchy   The generalized Cauchy (GC) family

$$k_{GC}(r; \ell_{\text{len}}, \alpha, \beta) \triangleq \left(1 + \left(\frac{r}{\ell_{\text{len}}}\right)^{\alpha}\right)^{-\beta/\alpha}, \qquad \alpha \in (0, 2], \quad \beta, \ell_{\text{len}} > 0 \tag{2.14}$$

was introduced by Gneiting and Schlather [72] to decouple the fractal dimension $D = d + 1 - \frac{\alpha}{2}$ (local erraticism) and the Hurst coefficient (long-range dependence) $H = 1 - \frac{\beta}{2}$ when $\beta \in (0, 1)$. Note that the GC family nests both the Cauchy ($\alpha = \beta = 2$ in (2.14)) and the RQ kernels. See [72] for further alternatives for matching the fractal dimension and the Hurst coefficient of f.

Power Exponential   The family of power exponential kernels are governed by the power hyperparameter $0 < \beta \le 2$:

$$k_{PE}(r; \beta, \ell_{\text{len}}) \triangleq \exp\left(-(r/\ell_{\text{len}})^{\beta}\right). \tag{2.15}$$

Special cases are Matérn-1/2 ($\beta = 1$) and SE ($\beta = 2$) kernels. Even though the direct shape of $k_{PE}(r)$ looks similar for varying β, the sample paths are highly sensitive to β. In fact the resulting process is never mean-square differentiable except when $\beta = 2$ [140]. Gramacy [79] states that $\beta < 2$ is good way to lower the condition number of $\mathbf{K}$, particularly in taking $\beta = 1.9$ for matrix inversion purposes, and when the smoothness of the fit is not important. The middle column of Fig. 2.3 shows that when $\beta = 1.9$, the sample paths have similar wiggliness as SE, but exhibit sharp turns. Examining the sample paths as β varies, they exhibit a similar rate of "long-term mean reversion" as the identical lengthscale should dictate. The major difference is in the short-term fluctuations, for example being extremely erratic for $\beta = 0.5$. Note that the previous Figs. 2.1 and 2.2 show the $\beta = 1$ and $\beta = 2$ cases.

Periodic Kernel   The *periodic kernel* is defined as

$$k_{per}(r; p, \ell_{\text{len}}) \triangleq \exp\left(-2\frac{\sin^2(\frac{\pi}{p}r)}{\ell_{\text{len}}^2}\right), \tag{2.16}$$

indexed by its *period length* p and lengthscale ℓ_{len} hyperparameters. This produces sample paths that are periodic, satisfying $f(x) = f(x \pm np)$ for $n \in \mathbb{Z}$. In this setting, the lengthscale is scaling $\sin^2(\cdot)$ rather than r. Still, ℓ_{len} controls the

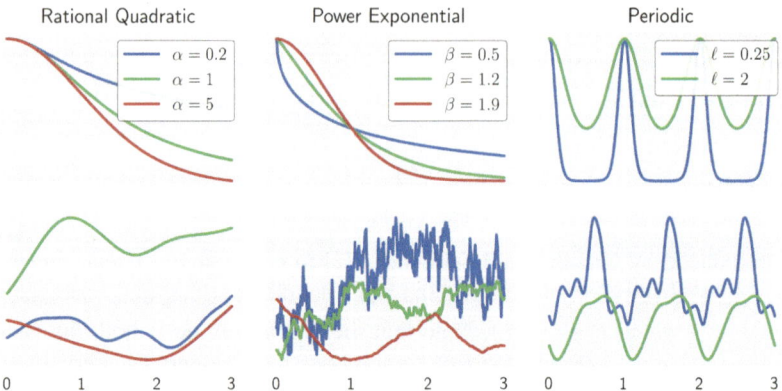

Fig. 2.3 Varying hyperparameters for stationary isotropic kernel families. Top row: $k(r)$ versus $r \in [0,3]$. Bottom row: corresponding sample paths. Rational quadratic (2.12) and power exponential (2.15) kernels use $\ell_{\text{len}} = 1$; the periodic kernel uses $p = 1$ in (2.16)

"wiggliness" of the resulting periodic function. This is seen in Fig. 2.3, where $\ell_{\text{len}} = 0.25$ results in a function with a more expressive pattern. Note however that since $p = 1$, both functions exactly produce $f(x) = f(x \pm 1) = f(x \pm 2) = \cdots$ for all x. The $\ell_{\text{len}} = 1$ case is seen in Fig. 2.1. Also note that $k(r)$ stays relatively large for all r when ℓ_{len} is also large. For example, the $\ell_{\text{len}} = 2$ case has $k(r)$ minimized when $r = 0.5, 1.5, 2.5, \ldots$, in which case $k(0.5) = 0.3679$.

Spectral Mixture Motivated by Bochner Theorem's 2.2 ability to represent stationary kernels through its spectral density, [168] develops two kernels, one being

$$k_I(r; \sigma, \lambda) \triangleq \exp\left(-2\pi^2\sigma^2 r^2\right) \cos\left(2\pi\lambda r\right), \tag{2.17}$$

which offers local periodicity and the ability to express negative prior correlation, and the other being the *spectral mixture (SM) kernel*

$$k_{SM}(\mathbf{h}; \boldsymbol{\theta}) \triangleq \sum_{q=1}^{Q} w_q \prod_{j=1}^{d} \exp\left(-2\pi h_j^2 v_q^{(j)}\right) \cos\left(2\pi |h_j| \mu_q^{(j)}\right), \tag{2.18}$$

which is a scale-location mixture of Gaussians, where Q, the w_q, $v_q^{(j)}$ and $\mu_q^{(j)}$, $j = 1, \ldots, d$ are all hyperparameters. Through Theorem 2.2, this kernel is able to approximate any stationary covariance kernel to arbitrary precision given Q large enough, since mixtures of Gaussians are dense in the set of all distribution functions [99]. This kernel is reported to perform well for pattern extrapolation with stationary data, for example see [170] to extrapolate a checkered pattern in image recognition, which is unable to be replicated with the SE or Matérn kernels.

White Noise   The *white noise kernel* is

$$k_{WH}(\mathbf{x}, \mathbf{x}') = 1_{\mathbf{x}=\mathbf{x}'}. \qquad (2.19)$$

k_{WH} allows to capture the concept of adding observation noise to a GP f with kernel k. Consider iid noise $\epsilon(\cdot)$ satisfying $\text{cov}(\epsilon(\mathbf{x}_i), \epsilon(\mathbf{x}_j)) = \sigma_\epsilon^2 \delta_{i=j}$. Then the covariance of $y(\cdot) = f(\cdot) + \epsilon(\cdot)$ is $k_f(\mathbf{x}_i, \mathbf{x}_j) + \sigma_\epsilon^2 k_{WH}(\mathbf{x}_i, \mathbf{x}_j)$ (assuming distinct inputs). In the case of the positive real line, one can consider fractional noise [72, 124] $k_{fWH}(x, x') = 1/2(|x - x' + 1|^{2H} - 2|x - x'|^{2H} + |x - x' - 1|^{2H}$, $H \in (0, 1)$.

Compact Kernels   A compactly supported kernel is zero over large distances: $k(r) = 0$ for $r > r_{max}$, where as usual $r = \|\mathbf{x} - \mathbf{x}'\|$. Compact kernels introduce zeros into the covariance matrix $\mathbf{K}$, bringing the potential to apply sparse linear algebra routines. For example, sparse matrix inversion methods might allow to speed up computation of $\mathbf{K}^{-1}$. This idea may be combined via covariance *tapering*: multiplying a regular kernel by a compactly supported one. Two common examples of compact kernels are Bohman and truncated power kernels:

$$k_B(r; r_{max}) \overset{\triangle}{=} \left(1 - \frac{r}{r_{max}}\right) \cos\left(\frac{\pi r}{r_{max}}\right) + \frac{1}{\pi} \sin\left(\frac{\pi r}{r_{max}}\right) \qquad (2.20)$$

$$k_{tp}(r; r_{max}) \overset{\triangle}{=} \left[1 - (\frac{r}{r_{max}})^\alpha\right]^\nu, \qquad 0 < \alpha < 2, \nu > \nu_m(\alpha). \qquad (2.21)$$

The term $\nu_m(\alpha)$ is difficult to calculate exactly, although Gneiting [69] provides upper bounds for a collection of α values. Note how r_{max} plays the role of controlling the sparsity (range of support of the kernel) and the lengthscale; it could be augmented with $r_{max} = r/\theta$ although identifiability becomes a potential issue. Compact kernels will inflate posterior variance since zeros in $\mathbf{K}$ make $\mathbf{K}^{-1}$ relatively larger; they also tend to make the posterior mean more wiggly [79, Chapter 9.1]. A typical solution [94] is multi-resolution, using a compact kernel to model local, short-range residuals after de-trending by a sufficiently flexible parametric global fit. See also [70].

2.2.2   Nonstationary Kernels

Polynomial Kernels   The basis for the family of polynomial kernels stems from the *dot product kernel* $k(\mathbf{x}, \mathbf{x}') = \mathbf{x}^\top \mathbf{x}'$. As opposed to stationary kernels, these measure how aligned $\mathbf{x}$ and $\mathbf{x}'$ are in $\mathbb{R}^d$, resulting in $k(\mathbf{x}, \mathbf{x}') = 0$ for orthogonal inputs and negative values for $\mathbf{x}'$ pointing in an opposite direction. The *inhomogeneous* version or *linear kernel* is $k_{\text{Lin}}(\mathbf{x}, \mathbf{x}') \overset{\triangle}{=} \sigma_0^2 + \mathbf{x}^\top \mathbf{x}'$ for the hyperparameter $\sigma_0 > 0$. This can be generalized to $k(\mathbf{x}, \mathbf{x}') = \sigma_0^2 + \mathbf{x}^\top A \mathbf{x}'$ for positive definite A.

The linear kernel connects Gaussian processes to Bayesian linear regression. Specifically, in applying a Bayesian framework to the simple linear regression line $f(x) = \beta_0 + \beta_1 x$ where $x \in \mathbb{R}$, if $\beta_0 \sim \mathcal{N}(0, \sigma_0^2)$, $\beta_1 \sim \mathcal{N}(0, 1)$, then we have $f \sim \mathcal{GP}(0, k_{\mathrm{Lin}})$. This can be verified through computing $\mathrm{cov}(f(x), f(x')) = \sigma_0^2 + x \cdot x'$. Scaling k_{Lin} yields a non-unit prior variance on β_1. The linear kernel thus allows to capture linear trends and can substitute for a linear mean function.

The *p-th degree polynomial kernel* is defined as

$$k_{\mathrm{poly}-p}(\mathbf{x}, \mathbf{x}'; \sigma_0) \stackrel{\triangle}{=} (\sigma_0^2 + \mathbf{x}^\top \mathbf{x}')^p, \qquad \sigma_0 \geq 0, \; p = 1, 2, \ldots, \tag{2.22}$$

which can be constructed through positive-integer powers maintaining positive definiteness (see Sect. 2.3). The case $\sigma_0 = 0$ yields the *homogeneous polynomial kernel* and $p = 1$ recovers the linear kernel.

Minimum Kernels The minimum kernel $k_{\mathrm{Min}}(x, x'; t_0) = t_0^2 + \min(x, x')$ corresponds to a Brownian motion process when $x \in \mathbb{R}_+$, where $t_0^2 = \mathrm{var}(f(0)) \geq 0$ is the initial prior variance. For discrete inputs $x \in \mathbb{Z}$, the Min kernel yields a Gaussian random walk. More generally, the *fractional Brownian motion kernel* over $\mathbb{R}^+$, corresponding to a fractional Brownian motion process, is $k_{\mathrm{fBM}}(x, x'; H) \stackrel{\triangle}{=} \frac{1}{2}(|x|^{2H} + |x'|^{2H} - |x - x'|^{2H})$ with Hurst coefficient $H \in (0, 1)$. The case $H = 1/2$ recovers the minimum kernel; if $H > 1/2$ ($H < 1/2$) then the increments are positively (resp. negatively) correlated. This is the only process over $\mathbb{R}_+$ that is Gaussian, has stationary increments, and is *self-similar*, i.e. $f(ax) \stackrel{d}{=} a^H f(x)$ for all $a \geq 0$, where the equality is in distribution [131]. The fact that it is self-similar means it satisfies the relationship $D + H = 2$, where D is its fractal dimension and H its Hurst coefficient.

Neural Tangent Kernel We mention the following connection between GPs and limits of fully-connected neural networks. Consider $f : \mathbb{R}^d \to \mathbb{R}$ modelled by a fully-connected artificial neural network with layers $l = 0, \ldots, L$. Thus $f(\cdot) = R_{L-1} \circ \cdots \circ R_0$, where $R_l = \sigma \circ A_l$ is the composition of the nonlinear activation function $\sigma(\cdot)$ and the affine $A_l(\mathbf{x}) = \frac{1}{\sqrt{n_l}} \mathbf{w}_l^\top \mathbf{x} + \beta$, and n_l is the size of layer l. Assume that each component of each $\mathbf{w}_l^\top \in \mathbb{R}^{n_l}$ is independent $\mathcal{N}(0, 1)$. [133] showed that as $n_l \to \infty$ for all l, the resulting $f(\cdot)$ tends to a Gaussian process with the *neural tangent kernel (NTK)* given by

$$k_{NTK}(\mathbf{x}, \mathbf{x}'; L + 1, \beta) = \Theta^{(L)}(\mathbf{x}, \mathbf{x}'), \tag{2.23}$$

where $\Theta^{(L)}$ is determined recursively and involves arc-cosine isotropic kernels of degree 0 and 1 [32]:

$$k_0(u) = \frac{1}{\pi}(\pi - \arccos(u)) \quad k_1(u) = \frac{1}{\pi}\left(u(\pi - \arccos(u)) + \sqrt{1 - u^2}\right).$$

Chen and Xu [30] proved that the NTK and Matérn-1/2 kernels have the same RKHS when $\mathbf{x} \in \mathbb{S}^{d-1}$, the unit sphere in d-dimensions, when using ReLU activation functions.

Non-separability
While all the aforementioned examples of multi-dimensional kernels have been separable, there are also many cases of non-separable kernels. Most of these originate in spatio-temporal statistics, where there is frequently a natural coupling between time- and space- correlation. For one example see (6.15) in Sect. 6.4 or [38, 157]. A recent survey of spatio-temporal kernels is in [136]. Such kernels could be especially relevant for financial applications when one of the input coordinates represents 'time'.

2.3 Kernel Composition and Engineering

Kernel composition in GPs refers to constructing complex kernels from simpler, "base" kernel families. This process exploits the algebraic properties of positive-definite functions, allowing the construction of new kernels via addition and multiplication. Specifically, if k_1 and k_2 are kernels and c_1, c_2 are positive real numbers, the linear combination $k(\mathbf{x}, \mathbf{x}') = c_1 k_1(\mathbf{x}, \mathbf{x}') + c_2 k_2(\mathbf{x}, \mathbf{x}')$ is also a kernel [19, 68, 148, 150]. This process mirrors the statistical concept of generalized additive models, where simpler models are summed to describe the overall response. Note that $c_1 k_1 + c_2 k_2$ is the kernel of the Gaussian process $c_1 f_1 + c_2 f_2$ obtained from independent GPs $f_1 \sim \mathcal{GP}(0, k_1)$ and $f_2 \sim \mathcal{GP}(0, k_2)$. Kernels are also closed under products, meaning $k(\mathbf{x}, \mathbf{x}') = k_1(\mathbf{x}, \mathbf{x}')k_2(\mathbf{x}, \mathbf{x}')$ is a kernel when k_1 and k_2 are. There is no GP analogue for products, however, as the multivariate normal distribution is not closed under (random vector) products.

Using inductive arguments, any polynomial $k(\mathbf{x}, \mathbf{x}') = c_0 + \sum_{i=1}^{I} c_i k_0(\mathbf{x}, \mathbf{x}')^i$ of a base kernel k_0 (with positive coefficients $c_i \geq 0$) is a kernel, cf. (2.22). As a corollary, $k = \exp(k_0)$, is also a kernel. Numerous additional constructions exist; the following all define a kernel $k(\mathbf{x}, \mathbf{x}')$ [68]:

$g(\mathbf{x})g(\mathbf{x}')$, for any $g : \mathbb{R}^d \to \mathbb{R}$;

$k_0(\phi(\mathbf{x}), \phi(\mathbf{x}'))$, for any $\phi : \mathbb{R}^d \to \mathbb{R}^p$ and kernel $k_0 : \mathbb{R}^p \times \mathbb{R}^p \to \mathbb{R}$;

$\mathbf{x}^\top A \mathbf{x}'$, for any positive definite matrix A;

$\frac{1}{4}\left[h(\mathbf{x}' + \mathbf{x}') - h(\mathbf{x} - \mathbf{x}')\right]$, for any nonnegative function $h : \mathbb{R}^d \to \mathbb{R}$.

The second notion above links to the input-scaling concept mentioned in Sect. 1.4.1 and can be extended to input *warping* [155]: in addition to projecting **x** linearly into the $[0, 1]^d$ cube as in (1.23), one considers $k(\mathbf{x}, \mathbf{x}') = \tilde{k}(\phi(\mathbf{x}), \phi(\mathbf{x}'))$ where $\phi : [0, 1]^d \rightarrow [0, 1]^d$ is a *learned* nonlinear transformation. For example, one can separately project each of the d coordinates of **x** using the BetaCDF transformation

$$\phi(x) = \int_0^x \frac{u^{\alpha-1}(1 - u)^{\beta-1}}{B(\alpha, \beta)} du$$

parametrized by $\alpha > 0, \beta > 0$ and $B(\alpha, \beta)$ being a corresponding normalizing constant. Each $\phi(\cdot; \alpha, \beta)$ is bijective from $[0, 1]$ to $[0, 1]$ and one can then augment $\alpha_1, \beta_1, \ldots, \alpha_d, \beta_d$ to the hyperparameters of k, optimizing them during the MLE step. This provides a hierarchical way of applying a non-stationary kernel and then a stationary one. Another warping function is the sigmoid $\phi(x) = \frac{1}{1+e^{-\alpha(x-\beta)}}$. Input *warping* is an effective technique to achieve stationarity in situations where there is a clear pattern of smoother response in some quadrant of $\mathcal{X}$ compared to others, though it requires more hyperparameters.

Locally Stationary Kernels
Silverman [151, 152] discusses *locally stationary kernels* of the form

$$k(\mathbf{x}, \mathbf{x}') = g\left(\frac{\mathbf{x} + \mathbf{x}'}{2}\right) k_S(\mathbf{x} - \mathbf{x}'), \qquad (2.24)$$

where g is a nonnegative function and k_S is a stationary kernel. The variance is determined by $k(\mathbf{x}, \mathbf{x}) = g(\mathbf{x})k_S(0) = g(\mathbf{x})$ if k_S is normalized so that $k_S(0) = 1$. Hence $g(\mathbf{x})$ describes the global structure, whereas $k_S(\mathbf{x} - \mathbf{x}')$ is invariant under shifts and thus describes the local structure. Using this, the heteroskedastic GP noise model can be recovered with $g(\mathbf{x})$ being the noise variance associated with **x** and $k_S(\mathbf{x}_i - \mathbf{x}'_j)$ being the white noise kernel.

2.4 Model Selection

Given the infinite variety of potential GP kernels one could utilize for a given task of fitting $(\mathbf{X}, \mathbf{y})$, how to compare their fitness? One solution is to rely the posterior likelihood of a kernel $k(\cdot, \cdot)$ given the data $p(k|\mathbf{y}) = p(\mathbf{y}|k)p(\mathbf{y})/p(k)$. Assuming a given (finite) space of kernels $\mathfrak{K}$ and a uniform (i.e. uninformed) prior that assigns all kernels equal weight, $p(k) = 1/|\mathfrak{K}|$, we see that $p(k|\mathbf{y}) \propto p(\mathbf{y}|k)$. However, the integral over hyperparameters $p(\mathbf{y}|k) = \int_\theta p(\mathbf{y}, \theta|k)d\theta$ is generally intractable. A common solution is to use the *Bayesian Information Criterion* (BIC) as an approximation, where $\text{BIC}(k) \approx -\log p(\mathbf{y}|k)$ is defined as [20]:

$$\text{BIC}(k) = -\ell_k(\hat{\theta}; \mathbf{y}) + \frac{|\hat{\theta}| \log(N)}{2}, \tag{2.25}$$

where $\ell_k(\theta|\mathbf{y}) = \log p(\mathbf{y}|k, \theta)$ is the log marginal likelihood of the N responses $\mathbf{y}$ evaluated at θ under a given kernel k, $\hat{\theta}$ is the maximizer (maximum marginal likelihood estimate) of $\ell_k(\theta|\mathbf{y})$, and $|\hat{\theta}|$ is the total number of kernel-specific hyperparameters. Note that $p(\mathbf{y}|k, \theta)$ is a MVN density, with mean and covariances governed by (1.8)–(1.9). The second term penalizes kernels that have more hyperparameters.

One can then assess the *relative* likelihood of two different kernels $k_1, k_2 \in \mathfrak{K}$ (still assuming a uniform prior over $\mathfrak{K}$) by computing the *Bayes Factor* (BF)

$$\text{BF}(k_1, k_2) \triangleq \frac{p(k_1|\mathbf{y})}{p(k_2|\mathbf{y})} \approx \exp\left(\text{BIC}(k_2) - \text{BIC}(k_1)\right). \tag{2.26}$$

The seminal work of [89] gives a table of *evidence categories* to determine a conclusion from Bayes factors. Typically a BF of more than 10 is viewed as *strong evidence* in favor of k_1 over k_2, and a BF over 100 is considered *decisive evidence* (hence BF less than 0.1 and 0.01 being evidence in the opposite direction).

Alternatively, cross-validation metrics can be used, particularly in prediction-based tasks. For example, the leave-one-out root mean-squared error (RMSE)

$$\text{RMSE}_{\text{LOO}} = \sqrt{\frac{1}{N} \sum_{i=1}^{N} (y_i - \mu_i)^2}, \tag{2.27}$$

can be computed efficiently (recall (1.24) for $\mu_i = \mathbb{E}[f(\mathbf{x}_i)|\mathcal{D}\backslash\{y_i\}]$).

Kernel Tree Search

Compositional kernels offer a rich and descriptive structure and in principle allow searching over the kernel space, solving the meta problem of kernel selection [50, 52]. For example, one can construct a tree-based structure for kernel composition, with base kernels representing tree leaf nodes and addition/multiplication operators representing internal tree nodes that combine sub-trees. Then, genetic algorithms (GA) can be used to explore kernel trees through mutation-selection operations, with the BIC metric used as the fitness criterion, see e.g. [117]. The GA probabilistically promotes exploration of the most fit kernels (ranked by their BIC after fitting to the given dataset $\mathcal{D}$), automating the discovery of the best-performing GP models.

2.4.1 Example: Kernel Fitness

We revisit the synthetic dataset $\mathcal{D}_{40}$ from Sect. 1.4.2 with the underlying response $f_0(x) = -0.1x^2 + \sin(3x)$ and iid noise $\epsilon \sim \mathcal{N}(0, 0.4^2)$. This time we fit the following three kernels

$$k_1 = \eta^2 k_{\text{Mat-5/2}}, \quad k_2 = \eta^2 k_{\text{Per}}, \quad k_3 = \eta_1^2 k_{\text{Poly2}} + \eta_2^2 k_{\text{Per}}.$$

The kernel k_1 represents a typical "safe" choice; it should do well in-sample (see Sect. 2.5 for a theoretical justification), but will suffer out-of-sample as it reverts to the prior $\hat{\mu}$ and will miss the non-stationary structure of f_0. The periodic kernel k_{Per} will be able to capture the term $\sin(3x)$ (which has period $2\pi/3 \approx 2.094$), but also mis-specified because $f_0(\cdot)$ is not fully periodic. The third kernel k_3 has the interpretation of the additive GP $f(x) = \eta_1^2 f_1(x) + \eta_2^2 f_2(x)$ capturing both the polynomial and periodic components.

Figure 2.4 illustrates the three posterior fits $m(x)$ with respective 95% uncertainty bands, $m(x) \pm 1.96s(x)$, and Table 2.1 provides goodness of fit metrics along with hyperparameter estimates. As expected, none of the models extrapolate well except for k_3 which has the added non-stationary component. Both periodic models k_2, k_3 approximately identify the period $p = 2.094$ (see Table 2.1), although

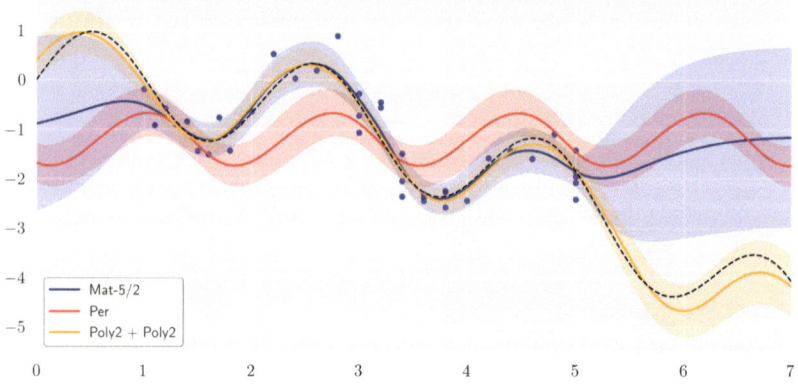

Fig. 2.4 Fitting the univariate response $f_0(x) = -0.1x^2 + \sin(3x)$ (dashed line) based on a dataset of 40 samples (shown as blue points) in the region $x \in [1, 5]$. We show three different fitted GP models along with their posterior 95% CIs; respective MLE hyperparameters are in Table 2.1

Table 2.1 Model performance for kernel choices k_1, k_2, and k_3 on $\mathcal{D}_{40}$, along with hyperparameter MLEs. Bolded values indicate the best for that metric

Kernel	$-l_k(\hat{\theta}; \mathbf{y})$	BIC	LOO	$|\hat{\theta}|$	$\hat{\eta}_1$	$\hat{\eta}_2$	$\hat{p}$	$\hat{\ell}_{\text{len}}$	$\hat{\sigma}_0$	$\hat{\sigma}_\epsilon$	$\hat{\mu}$
$\eta^2 k_{\text{M5/2}}$	-0.771	**38.23**	0.436	4	0.859	–	–	0.634	–	0.379	-1.223
$\eta^2 k_{\text{Per}}$	-1.229	58.39	0.819	5	0.257	–	1.717	1.445	–	0.767	-1.196
$\eta_1^2 k_{\text{Poly2}} + \eta_2^2 k_{\text{Per}}$	-0.672	39.79	**0.397**	7	0.011	1.227	2.166	4.103	0.0935	0.369	0.0324

$f_0(\cdot)$ not being fully periodic poses severe issues for the k_2 both in- and out-of sample. This is further verified by its BIC and RMSE$_{\text{LOO}}$ scores. The third model appears best visually and on the basis of RMSE$_{\text{LOO}}$ (out-of-sample prediction). Surprisingly, k_1 has the best BIC, though comparing Bayes factors $\widehat{\text{BF}}(k_1, k_3) = \exp(39.79 - 38.23) = 4.7588$ which is in the category of "substantial", but not "strong" evidence [89]. Moreover, k_3 is a complex model for the relatively low data size $N = 40$ and adds three additional hyperparameters (the second outputscale η_2 plus two hyperparameters inside k_{Per}) compared to $k_{\text{M5/2}}$. While k_3 offers a predictive improvement, this does not directly translate to *in-sample* distributional fit, which is what BIC proxies. If we measured predictive accuracy say for $x > 5$, then k_3 is likely to come out on top.

Polynomial Priors: Mean Versus Covariance

Should the polynomial component be placed in the prior mean function or in the covariance kernel? Relating to this example, it is not immediately obvious if k_2 equipped with prior mean $\mu(x) = \beta_0 + \beta_1 x + \beta_2 x^2$ is "better" than k_3. Ultimately, they produce the same quadratic extrapolation, and often perform similarly in-sample. Placing this assumption on the prior mean is generally the recommendation (e.g. [79]), with reasons including an explicitly estimated parametric mean function and less bloated compositional kernel. Note however that except when clear prior knowledge is known, higher degree polynomial extrapolation is generally dangerous, so we recommend limiting the prior mean to a linear function.

2.5   Convergence and Universal Approximation

Given the non-parametric nature of GPs, a natural question is whether a GP f fitted to $\mathcal{D}_N$ "converges" to the ground truth f_0 as $N \to \infty$. We distill the arguments in [91] who show that relatively common assumptions indeed lead to such convergence, along with convergence rates. To provide a concrete setting, consider $\mathcal{X} = [0, 1]^d$, assume that f_0 is Hölder continuous with order $\beta > d/2$ and belongs to the Sobolev space $H^\beta([0, 1]^d)$, and that $\mathcal{D}_N$ contains $\mathbf{x}$'s sampled according to a given distribution $\mathbb{P}_X$ bounded away from 0 and ∞. Suppose that $f \sim \mathcal{GP}(0, k_\nu)$, where k_ν is a Matérn-ν kernel. Then,

$$\mathbb{E}\left[\int \|f - f_0\|_{L^2(\mathbb{P}_X)}^2 d\Pi_N(f|\mathcal{D}_N)\right] = O(N^{-2\min(\nu,\beta)/(2\nu+d)}), \qquad (2.28)$$

where $\Pi_N(f|\mathcal{D}_N)$ is the posterior distribution of $f|\mathcal{D}_N$, and the outer expectation is over $\mathcal{D}_N|f$ (i.e. over $\mathbb{P}_X$ and the iid ϵ).

Note 2.3 *There are a few key remarks to make:*

- The best rate is attained when $v = \beta$, resulting in convergence rate of $O(N^{-2\beta/(2\beta+d)})$. In this case, the smoothness of the GP is the same as the underlying f_0, and $\frac{2\beta}{2\beta+d}$ is the minimax-optimal rate for regression of a function in $W^\beta([0, 1]^d)$.
- Since $\|m_N - f_0\|^2_{L^2(\mathbb{P}_X)} \leq \int \|f - f_0\|^2_{L^2(\mathbb{P}_X)} d\Pi_N(f|\mathcal{D}_N)$, where m_N is the posterior mean (now emphasizing dependence on N), it follows that the same convergence rate in (2.28) applies to the posterior mean.
- This result holds for any GP whose kernel's RKHS is a Sobolev space, since the Matérn kernel RKHS is equivalent to $H^{v+d/2}([0, 1]^d)$.
- The assumption that $\mathbb{P}_X$ is bounded away from 0 and ∞ ensures that the sampling is done "uniformly enough" on $[0, 1]^d$.
- GPs with SE kernels are also known to be universal approximators to within an arbitrarily small ϵ-band of any continuous f_0 [130].

Note that the assumptions on the underlying f_0 are relatively mild, encompassing a wide range of function classes The weak derivatives required by Sobolev spaces allow for functions with certain types of singularities or discontinuities, making the assumption inclusive of a variety of real-world scenarios where functions may not be perfectly smooth. Hence even non-smooth or irregular financial data can be modeled effectively, making the Matérn class of kernels a "safe" choice when little is known about the underlying f_0.

Curse of Dimensionality and Posterior Standard Deviation
In a different direction, consider the noiseless case $\sigma_\epsilon \equiv 0$. For a given input **x**, suppose that $\mathbf{x}_1, \ldots, \mathbf{x}_N$ are equally-spaced grid points in a ball of radius ρ around **x**. Then the posterior standard deviation is $s_N(\mathbf{x}) = O(N^{-2\rho/d})$. This explicitly reveals the curse of dimensionality, as the above requirement to achieve a target width for the posterior interval $m_N(x) \pm s_N(x)$ increases exponentially in d. Kanagawa et al. [91] discusses a more general result regarding a more arbitrary fill of points.

2.6 Connections with SDEs and Other Processes

Readers of this book are probably familiar with the theory of stochastic differential equations (SDE). SDEs offer another interpretation for one-dimensional Gaussian processes. To match the standard SDE notation, we consider a $\mathbb{R}$-valued GP f, indexed by "time" t. In what follows, we recall the connection between f being a GP and f being a solution of an SDE with a Brownian motion driver. The primary intuition is that a linear SDE with Markov coefficients has a Gaussian transition density, and hence has Gaussian fdd.

Let $(W_t)_{t \geq 0}$ be a Wiener process and recall the Ornstein-Uhlenbeck SDE

$$df(t) = -\kappa f(t)dt + \sigma \, dW_t \tag{2.29}$$

Starting with an initial distribution $f(0) \sim \mathcal{N}(0, \frac{\sigma^2}{2\kappa})$ we obtain a stationary process with $\mathbb{E}[f(t)] = 0$ for all t, and $\text{cov}(f(t), f(t')) = \frac{\sigma^2}{2\kappa} \exp(\frac{-|t-t'|}{\kappa})$. We recognize that the latter is the covariance of the Matérn-1/2 GP (2.9). The other SDE connection we have already encountered is $df(t) = \sigma dW_t$, i.e. $f(t) = \sigma W_t$ which corresponds to a nonstationary GP with the Min kernel.

One may obtain [83] exact linear-SDE representations for the Matérn-ν GPs with $\nu + 1/2 \in \mathbb{Z}$, by creating an auxiliary chain of $\nu + 1/2$ additional coordinates where only the last coordinate has a non-zero diffusion term. To illustrate, a GP f with the Matérn-3/2 kernel k_{M32}, lengthscale ℓ_{len} and outputscale η^2 admits the representation [147, Example 12.7] $f(t) = X_t^{(1)}$ where:

$$\begin{cases} dX_t^{(1)} = X_t^{(2)} dt; \\ dX_t^{(2)} = (-\lambda^2 X_t^{(1)} - 2\lambda X_t^{(2)})dt + 2\sigma\lambda^{3/2} dW_t, \end{cases} \tag{2.30}$$

where $\lambda = \sqrt{3}/\ell_{\text{len}}$ and the initial distributions are Gaussian $X_0^{(1)} \sim \mathcal{N}(0, \eta^2)$, $X_0^{(2)} \sim \mathcal{N}(0, \lambda^2 \eta^2)$. Observe that $X^{(2)}$ is understood as the temporal *derivative* of f, or alternatively, f is the time-integral of $X^{(2)}$. Similarly, for Matérn-5/2 (2.11) we have, using matrix notation, $f(t) = X_t^{(1)}$,

$$d \begin{pmatrix} X_t^{(1)} \\ X_t^{(2)} \\ X_t^{(3)} \end{pmatrix} = \begin{pmatrix} 0 & 1 & 0 \\ 0 & 0 & 1 \\ -\lambda^3 & -3\lambda^2 & -3\lambda \end{pmatrix} \begin{pmatrix} X_t^{(1)} \\ X_t^{(2)} \\ X_t^{(3)} \end{pmatrix} dt + \frac{4\sigma\lambda^{5/2}}{\sqrt{3}} \begin{pmatrix} 0 \\ 0 \\ 1 \end{pmatrix} dW_t.$$

Such SDE representations offer a further connection between the GP posterior equations and the theory of stochastic smoothing, i.e. computing conditional expectations of $f(t)$ given observations at time $t_1, \ldots, t_N$ [147].

Remark 2.4 The SE kernel, being a limit of Matérn-ν as $\nu \rightarrow \infty$, requires an infinite chain of auxiliary $X^{(i)}$ processes, but can be approximated through truncation. Similar approximations exist for the rational quadratic k_{RQ} and periodic k_{per} kernels.

Going in the opposite direction, one may use SDEs to generate additional kernels. For example, taking $f(\cdot)$ to be the integrated Brownian motion process, $df_t = (\sigma W_t)dt$, and computing its covariance function yields the GP kernel

$$k(t, t') = \sigma \left(\frac{\min(t, t')^3}{3} + |t - t'| \frac{\min(t, t')^2}{2} \right).$$

The above construction may be generalized via *autoregressive* GPs. A *p*-order autoregressive GP is stationary Gaussian process $\{f(x) : x \in \mathbb{R}_+\}$ satisfying

$$\partial_x^p f(x) + a_{p-1} \partial_x^{p-1} f(x) + \cdots + a_0 f(x) = b_0 Z(x), \qquad (2.31)$$

where $\partial_x^p f$ is the *p*-th derivative of f with respect to x and Z is the *white noise* process with kernel $k_{WH}(x, x') = \delta(x - x')$. The intuition of (2.31) is to bridge to the Markov property that is a key feature of SDE solutions.

Utilizing a Fourier transform approach, [140, Appendix B] shows that the power spectrum of f satisfying (2.31) is

$$S(u) = \frac{b_0^2}{|\sum_{k=0}^{p} a_k (2\pi i u)^k|^2}.$$

Then, Bochner's Theorem 2.2 and the corresponding Fourier inversion gives the covariance kernel of f. For example, with $p = 1$, we have $S(u) = \frac{b_0^2}{(2\pi u)^2 + a_0^2}$ or

$$k(x, x') = \frac{b_0^2}{2a_0} e^{-a_0|x-x'|}, \qquad (2.32)$$

which matches the Matérn-1/2 covariance (2.9) for $a_0 = \frac{1}{\ell_{\text{len}}}$ and $b_0 = \frac{2}{\ell_{\text{len}}}$. More generally, a Matérn-$(p + 1/2)$ process satisfies a *p*-th order autoregressive scheme, which in turn confirms that its sample paths are *p*-times differentiable. Intuitively, a process becomes rougher with each differentiation and (2.31) implies that $\partial_x^p f$ resembles the white noise process, which is not mean-square continuous.

As another example, consider a GP f satisfying the second order autoregressive scheme

$$f''(x) + 2\alpha f'(x) + \gamma^2 f(x) = Z(x). \qquad (2.33)$$

If $\omega^2 = \gamma^2 - \alpha^2 > 0$, then the covariance kernel of f is the AR2 kernel [134]:

$$k(x, x') = \frac{\exp(-\alpha|x-x'|)}{4\alpha\gamma^2} \left\{ \cos\omega|x-x'| + \frac{\alpha}{\omega} \sin\omega|x-x'| \right\}. \qquad (2.34)$$

Similarly, integrating against (fractional) Brownian motions yields Gaussian processes when the integrands are deterministic. Consider

$$f(t) = \int_0^t a_s dW_s, \qquad f_H(t) = \int_0^t a_s dW_s^H, \qquad (2.35)$$

where W^H is a fractional Brownian motion. Both of these define mean zero GPs. A routine calculation reveals the respective covariance kernels to be $k(s, t) =$

$\int_0^{s \wedge t} a_u^2 du$ and $k_H(s, t) = H(2H - 1) \int_0^s \int_0^t a_u a_v |u - v|^{2H-2} du dv$. A special case is the *Gaussian Volterra* process [132], which takes the form

$$f(t) = \int_0^t \mathcal{K}(t, s) \, dW_s, \tag{2.36}$$

where W is a standard Brownian motion and $\mathcal{K}$ is a *Volterra kernel* meaning $\mathcal{K}(t, s) = 0$ for $s > t$. When $\int_0^T \int_0^T \mathcal{K}(t, s) \, ds dt < \infty$, we have that Eq. (2.36) defines a GP over $[0, T]$. In this case, f has mean zero and covariance $k(t, s) = \int_0^{t \wedge s} \mathcal{K}(t, u) \mathcal{K}(s, u) du$. Thus, the Volterra kernel $\mathcal{K}$ is a sort of "square root" of the resulting covariance kernel k.

Non-continuous Inputs
While throughout this Chapter we assume that the inputs $\mathbf{x}$ are continuous, there is no mathematical difficulty in working with discrete $\mathcal{X}$. For example, discrete autoregressive processes naturally link to time-series modeling. Moreover, one may consider categorical inputs [66], which are usually translated into a discrete $\mathcal{X}$ through one-hot encoding: assigning a 0/1 flag for each possible category, augmenting these flags to $\mathbf{x}$.

Further Reading

The material is scattered throughout many sources. We mainly borrow from the monograph of Rasmussen and Williams [140] which has been the standard reference for the past 20 years and provides a unified treatment of Gaussian Markov Processes (see their Appendix B). For broader theory and required mathematical machinery to make arguments rigorous, see [3, 134] and [141]. The seminal work on providing a full theory of reproducing kernels appears in Aronszajn [8].

The article [91] bridges gaps with GP models and frequentist kernel methods, broadening connections with kernel ridge regression, the theory of RKHS, and more. The recent monograph by Sarkka and Solin [147] provides rigorous treatment regarding the connections between SDEs and GPs. We also mention the kernel "cookbook" [51] which provides a starting point for combining base kernels in bespoke ways.

This Chapter is accompanied by a `Python Jupyter` notebook illustrating fitting of different GP kernels and prior mean functions to a synthetic one-dimensional dataset, with code that reproduces Figs. 2.1, 2.2, 2.3, and 2.4.

Chapter 3
Advanced GP Modeling Topics

This chapter discusses a variety of ways to extend or modify the core GP model to make it better suited for concrete applications. Heteroskedastic GPs (Sect. 3.1) are essential in many financial applications to capture input-dependent noise. Alternative likelihoods (Sect. 3.2) tackle non-Gaussian noise. Multi-output GPs (Sect. 3.3) provide a framework to capture correlation across multiple surrogates. Section 3.4 surveys localization methods that can be used to either combat nonstationarity or reduce runtime. Further (exact) computational improvements are given through the GP updating equations in Sect. 3.5, useful in the setting of "streaming data."

3.1 Heteroskedastic GPs

Input-dependent noise is ubiquitous when simulating stochastic models in finance. The cause can be as simple as state-dependent (e.g. stochastic or local) volatility, or the nonlinear payoffs involved in carrying out Monte Carlo estimation of expected values. Many statistical frameworks, such as the classical multivariate linear regression, rely heavily on the assumption of constant, known, Gaussian noise. GP model fitting is also prone to some challenges in heteroskedastic settings. Indeed, dramatic variation in signal-to-noise ratios as input varies distorts the log-likelihood function introduced in Sect. 1.4.1.

We revisit the GP observation model (1.5) for data outputs $y(\mathbf{x})$, generalizing to explicitly heteroskedastic noise:

$$y(\mathbf{x}_i) = f(\mathbf{x}_i) + \varepsilon_i, \qquad \varepsilon_i \sim \mathcal{N}(0, r(\mathbf{x}_i)), \tag{3.1}$$

where $r(\mathbf{x}_i)$ is the state-dependent variance. In matrix form, assuming zero prior mean and the training set $(\mathbf{X}, \mathbf{y})$ of size N, we have $\mathbf{y} \sim \mathcal{N}(\mathbf{0}, \mathbf{K}_f + \mathbf{\Sigma})$,

© The Author(s), under exclusive license to Springer Nature Switzerland AG 2025   49
M. Ludkovski, J. Risk, *Gaussian Process Models for Quantitative Finance*,
SpringerBriefs in Quantitative Finance,
https://doi.org/10.1007/978-3-031-80874-6_3

where $\mathbf{K}$ is as usual the $N \times N$ matrix with i, j coordinates $k(\mathbf{x}_i, \mathbf{x}_j)$, and $\mathbf{\Sigma} = \mathrm{diag}(r(\mathbf{x}_1), \ldots, r(\mathbf{x}_N))$ is the noise matrix.

The posterior $f_*(\mathbf{x}_*)|\mathcal{D} \sim \mathcal{N}(m(\mathbf{x}_*), s^2(\mathbf{x}_*))$ is Gaussian with the familiar parameters (cf. (1.8)–(1.9))

$$m(\mathbf{x}_*) = \mathbf{k}(\mathbf{x}_*)^\top (\mathbf{K}_f + \mathbf{\Sigma})^{-1} \mathbf{y}, \quad \text{where } \mathbf{k}(\mathbf{x}_*) = (k(\mathbf{x}_*, \mathbf{x}_1), \ldots, k(\mathbf{x}_*, \mathbf{x}_N))^\top; \tag{3.2}$$

$$s^2(\mathbf{x}_*) = k(\mathbf{x}_*, \mathbf{x}_*) + r(\mathbf{x}_*) - \mathbf{k}(\mathbf{x}_*)^\top (\mathbf{K}_f + \mathbf{\Sigma})^{-1} \mathbf{k}(\mathbf{x}_*). \tag{3.3}$$

The use of (3.2) requires knowing the noise matrix $\mathbf{\Sigma}$. If this is not available, estimating the underlying $r(\cdot)$ is non-trivial, since it is never observed directly and hence one must disentangle signal from noise, and then estimate the level of the noise. One approach to facilitate this task is to *replicate* inputs. Replicates at $\mathbf{x}$ allow to directly see the variability of the corresponding outputs which by definition are identically distributed with variance $r(\mathbf{x})$. Note that in (3.1) $\mathrm{cov}(\varepsilon_i, \varepsilon_j) = 0$ even if $\mathbf{x}_i = \mathbf{x}_j$ is the same location (but distinct samples).

A further advantage of replication is data-size reduction via switching from the full-N size of the data to the unique-n number of unique input locations. Let $\bar{\mathbf{x}}_i$, $i = 1, \ldots, n$ represent the $n \ll N$ unique input locations, ordering the data as

$$\mathbf{X} = (\underbrace{\bar{\mathbf{x}}_1, \ldots, \bar{\mathbf{x}}_1}_{a_1 \text{ times}}, \ldots, \underbrace{\bar{\mathbf{x}}_n, \ldots, \bar{\mathbf{x}}_n}_{a_n \text{ times}})^\top \tag{3.4}$$

where each input is repeated a_i times and $\sum_{i=1}^n a_i = N$. Stack the observations $\mathbf{y}$ in the same order and let $y_i^{(j)}$, $j = 1, \ldots, a_i$ be the $a_i \geq 1$ replicates observed at $\bar{\mathbf{x}}_i$. Then let $\bar{\mathbf{y}} = (\bar{y}_1, \ldots, \bar{y}_n)^\top$ collect averages of replicates,

$$\bar{y}_i = \frac{1}{a_i} \sum_{j=1}^{a_i} y_i^{(j)}.$$

Due to the independence of $y_i^{(j)}$, the variance of $\bar{y}_i$ is reduced proportionally to a_i^{-1}, $\mathrm{var}(\bar{y}_i) = \frac{\Sigma_{ii}}{a_i}$. Note that we allow the number of replicates $a_i \geq 1$ vary site-to-site, subsuming the unreplicated setting.

Let $\mathbf{\Sigma}_n = \mathrm{diag}(r(\bar{\mathbf{x}}_1), \ldots, r(\bar{\mathbf{x}}_n))$, $\mathbf{A}_n = \mathrm{diag}(a_1, \ldots, a_n)$ and $\mathbf{U}$ be the $N \times n$ block matrix $\mathbf{U} \stackrel{\triangle}{=} \mathrm{diag}(\mathbf{1}_{a_1,1}, \ldots, \mathbf{1}_{a_n,1})$, where $\mathbf{1}_{k,l}$ is $k \times l$ matrix filled with ones. Then we have $\mathbf{X} = \mathbf{U}\bar{\mathbf{X}}$, and similarly, $\mathbf{K} = \mathbf{U}\mathbf{K}_n\mathbf{U}^\top$, where $\mathbf{K}_n = (k(\bar{\mathbf{x}}_i, \bar{\mathbf{x}}_j))_{1 \leq i, j \leq n}$ [22]. Thus, $\mathbf{U}$ is the mapping from the full N to the small-n dimension. For example, $\mathbf{U}^\top \mathbf{\Sigma} \mathbf{U} = \mathbf{A}_n^{-1} \mathbf{\Sigma}_n = \mathrm{var}(\bar{\mathbf{y}})$.

Let $\mathbf{k}_n(\mathbf{x}_*) \stackrel{\triangle}{=} (k(\mathbf{x}_*, \bar{\mathbf{x}}_1), \ldots, k(\mathbf{x}_*, \bar{\mathbf{x}}_n))^\top$. Then we have [7] the unique-n predictive equations

$$m_n(\mathbf{x}_*) = \mathbf{k}_n(\mathbf{x}_*)^\top (\mathbf{K}_n + \mathbf{A}_n^{-1}\mathbf{\Sigma}_n)^{-1}\bar{\mathbf{y}}, \tag{3.5}$$

$$s_n^2(\mathbf{x}_*) = k(\mathbf{x}_*, \mathbf{x}_*) + r(\mathbf{x}_*) - \mathbf{k}_n(\mathbf{x}_*)^\top (\mathbf{K}_n + \mathbf{A}_n^{-1}\mathbf{\Sigma}_n)^{-1}\mathbf{k}_n(\mathbf{x}_*). \qquad (3.6)$$

Observe the full analogy between Eqs. (3.5)–(3.6) and the earlier (3.2), which establishes that the GP posterior mean and variance, calculated on the average responses at replicates and with covariances calculated only at the unique design sites, is identical to the original GP equations that ignore replication. In turn these identities imply computational savings: one can handle $N \gg n$ training inputs in $O(n^3)$ time (by only utilizing the unique-n observations in $\bar{\mathbf{y}}$) if many of them are replicates.

A well-known way to estimate $\mathbf{\Sigma}$ is *stochastic kriging* (SK) due to Ankenman et al. [7]. SK leverages replication to estimate $r(\mathbf{x}_i)$ through the standard empirical plug-in (cf. (3.7) below), replacing $\mathbf{\Sigma}_n$ in (3.5) with

$$\widehat{\mathbf{\Sigma}}_n = \mathrm{diag}(\hat{\sigma}_1^2, \dots, \hat{\sigma}_n^2), \quad \text{where} \quad \hat{\sigma}_i^2 \overset{\triangle}{=} \frac{1}{a_i - 1} \sum_{j=1}^{a_i} (y_i^{(j)} - \bar{y}_i)^2. \qquad (3.7)$$

When $a_i \gg 1$ ([7] recommend $a_i \geq 10$), the resulting $\widehat{m}_n(\mathbf{x})$ is still unbiased. However, an important caveat is that SK is zero-bias only for fixed GP hyperparameters. In the case that the latter are learned, for instance via MLE, the bias re-appears. Indeed, while the little-n predictive equations with known $\mathbf{\Sigma}$ are exact, the SK likelihood using pre-averaged values

$$\ell_n^{SK}(\boldsymbol{\theta}|\mathbf{y}) = \mathrm{Const} - \frac{n}{2}\log(2\pi) - \frac{1}{2}\log|\mathbf{K}_n + \mathbf{A}_n^{-1}\widehat{\mathbf{\Sigma}}_n| - \frac{1}{2}\bar{\mathbf{y}}^\top(\mathbf{K}_n + \mathbf{A}_n^{-1}\widehat{\mathbf{\Sigma}}_n)^{-1}\bar{\mathbf{y}}$$
$$(3.8)$$

is an approximation, affecting the inference of the GP hyperparameters. Moreover, since $\hat{\sigma}(\cdot)$ is only estimated at $\mathbf{x}_1, \dots, \mathbf{x}_n$ there is no immediate way to predict intrinsic variance at out-of-sample $\mathbf{x}_*$'s. Ankenman et al. [7] propose a second, separately estimated, (no-noise) GP f_r for $r(\cdot)$ trained on the $(\mathbf{x}_i, \hat{\sigma}_i^2)$ pairs to obtain a prediction $m_r(\mathbf{x}_*) = \mathbb{E}[r(\mathbf{x}_*)|\mathcal{D}]$.

Alternatively, Binois et al. [22] jointly infer a GP f_r for $r(\cdot)$ along with the GP for $f(\cdot)$ by exploiting the structure of replicated designs for GPs via the Woodbury identities (A.2). The connection to GPs is to re-parametrize through the correlation matrix $\mathbf{K}_f = \eta^2 \mathbf{C}_f$, $\mathbf{K}_n = \eta^2 \mathbf{C}_n$, $\mathbf{\Sigma} = \eta^2 \mathbf{\Lambda}$, factoring out the process variance η^2 and then set $\mathbf{D} = \mathbf{\Sigma} = \eta^2 \mathbf{\Lambda}$ and $\mathbf{V} = \mathbf{U}^\top$, which allows the computation of the inverse and determinant of $(\mathbf{K}_f + \mathbf{\Sigma})$ without ever building the full-N matrices.

Lemma 3.1 *Let* $\mathbf{U}_n \overset{\triangle}{=} \mathbf{C}_n + \mathbf{A}_n^{-1}\mathbf{\Lambda}_n$ *and* $\mathbf{\Lambda}_n = \mathrm{diag}(\lambda_1, \dots, \lambda_n)$. *The following unique-n identity holds for the full-N expression for the conditional log likelihood,*

$$\ell(\boldsymbol{\theta}|\mathbf{y}) = Const - \frac{N}{2}\log\hat{\eta}^2 - \frac{1}{2}\sum_{i=1}^{n}\left[(a_i-1)\log\lambda_i + \log a_i\right] - \frac{1}{2}\log|\mathbf{U}_n|,$$

(3.9)

where $\qquad \hat{\eta}^2 \triangleq \frac{1}{N}\left(\mathbf{y}^\top\boldsymbol{\Lambda}^{-1}\mathbf{y} - \bar{\mathbf{y}}^\top\mathbf{A}_n\boldsymbol{\Lambda}_n^{-1}\bar{\mathbf{y}} + \bar{\mathbf{y}}^\top\mathbf{U}_n^{-1}\bar{\mathbf{y}}\right).$ (3.10)

The proof of the lemma is based on the following key computations:

$$\mathbf{y}^\top(\mathbf{C}_f + \boldsymbol{\Lambda})^{-1}\mathbf{y} = \mathbf{y}^\top\boldsymbol{\Lambda}^{-1}\mathbf{y} - \bar{\mathbf{y}}^\top\mathbf{A}_n\boldsymbol{\Lambda}_n^{-1}\bar{\mathbf{y}} + \bar{\mathbf{y}}^\top(\mathbf{C}_n + \mathbf{A}_n^{-1}\boldsymbol{\Lambda}_n)^{-1}\bar{\mathbf{y}};$$ (3.11)

$$\log|\mathbf{C} + \boldsymbol{\Lambda}| = \log|\mathbf{C}_n + \mathbf{A}_n^{-1}\boldsymbol{\Lambda}_n| + \sum_{i=1}^{n}\left[(a_i-1)\log\lambda_i + \log a_i\right].$$ (3.12)

The pre-averaged log-likelihood (3.8) used by SK would estimate the process variance using unique-n calculations based on $\bar{\mathbf{y}}$ as $\hat{\eta}_n^2 = n^{-1}\bar{\mathbf{y}}^\top\mathbf{U}_n^{-1}\bar{\mathbf{y}}$, while (3.10) gives

$$\hat{\eta}^2 = \frac{1}{N}\left(\mathbf{y}^\top\boldsymbol{\Lambda}^{-1}\mathbf{y} - \bar{\mathbf{y}}^\top\mathbf{A}_n\boldsymbol{\Lambda}_n^{-1}\bar{\mathbf{y}} + n\hat{\eta}_n^2\right) = \frac{1}{N}\sum_{i=1}^{n}\frac{a_i}{\lambda_i}\check{\sigma}_i^2 + \frac{n}{N}\hat{\eta}_n^2,$$

where $\check{\sigma}_i^2 = \frac{1}{a_i}\sum_{j=1}^{a_i}(y_i^{(j)} - \bar{y}_i)^2$, i.e., the bias un-adjusted estimate of $r(\mathbf{x}_i)$ based on $\{y_i^{(j)}\}_{j=1}^{a_i}$.

As a second step, the entries $\lambda_1, \ldots, \lambda_n$ of $\boldsymbol{\Lambda}_n$ are recast as posterior log-means of a second GP $f_r \sim \mathcal{GP}(0, k_r)$ that regularizes the unobserved noise values to capture the fact that the estimates of $r(\mathbf{x})$ based on $(\bar{\mathbf{X}}, \bar{\mathbf{y}})$ are themselves noisy [74]. In vector form, one sets

$$\log\boldsymbol{\Lambda}_n = \mathbf{C}_{(g)}(\mathbf{C}_{(g)} + g\mathbf{A}_n^{-1})^{-1}\boldsymbol{\Delta}_n$$ (3.13)

where $\mathbf{C}_{(g)}$ is the covariance matrix of f_r and g is its observation noise, appropriately scaled by the replication counts. Such smoothing allows to handle $a_i = 1$: even if there is no replication one borrows information from other $\mathbf{x}_j$'s to infer $r(\mathbf{x}_i)$. The $\boldsymbol{\Delta}_n$ are inferred through MLE together with the hyperparameters of k_r.

3.2   Alternative Likelihood Functions

A related feature that often arises when using GPs in practice is the fact that the observation noise $\epsilon(\mathbf{x})$ is often non-Gaussian. By construction, the noise is assumed to be centered, and non-Gaussianity is typically manifested in terms of heavier tails. In turn, mis-specified noise may lead to poor fit; for instance the model may mistake large noise realizations for signal, and overfit to the respective outlier observations. In this section we discuss extensions of the framework to handle such situations, which reduce to replacing the Gaussian likelihood (1.19) with a non-Gaussian version. The main challenge is that the marginal likelihood becomes analytically intractable and requires approximation. Indeed, Gaussian observation noise is the unique case to achieve Bayesian conjugacy between the GP prior and the posterior.

3.2.1 Gaussian Process Regression with Student t-Noise

A first alternative is to assume that the noise ϵ_i at input $\mathbf{x}_i$ has a Student-t distribution [90, 163]. In particular, this may lead to better fits when the noise is heavy-tailed, making inference more resistant to outliers. Specifically, consider i.i.d. noise $\epsilon_i \sim t_\nu(\sigma_\epsilon^2)$ that is t-distributed with variance σ_ϵ^2 and $\nu > 2$ degrees of freedom (the latter is treated as another hyperparameter). Using the Gamma function $\Gamma(\cdot)$, the likelihood of observing $\mathbf{y}$ conditional on the latent values $\mathbf{f}$ becomes

$$p_{t\text{GP}}(\mathbf{y}|\mathbf{f}) = \prod_{i=1}^{n} \frac{\Gamma((\nu+1)/2)}{\Gamma(\nu/2)\sqrt{\nu\pi}\sigma_\epsilon} \times \left(1 + \frac{(y_i - f_i)^2}{\nu\sigma_\epsilon^2}\right)^{-(\nu+1)/2}. \tag{3.14}$$

Integrating (3.14) against the Gaussian prior $f \sim \mathcal{GP}(0, k)$ is intractable. The most common workaround is to use the Laplace approximation (LP) method [163] to calculate the posterior. A second-order Taylor expansion of $\log p_{t\text{GP}}(\mathbf{f}|\mathbf{y})$ around its mode, $\tilde{\mathbf{m}}_{t\text{GP}} \triangleq \arg\max_{\mathbf{m}} p_{t\text{GP}}(\mathbf{f}|\mathbf{y})$, gives a Gaussian approximation

$$p_{t\text{GP}}(\mathbf{f}|\mathbf{y}) \approx q_{t\text{GP}}(\mathbf{f}|\mathbf{y}) \triangleq \mathcal{N}\left(\tilde{\mathbf{m}}_{t\text{GP}}, \boldsymbol{\Sigma}_{t\text{GP}}^{-1}\right), \tag{3.15}$$

where $\boldsymbol{\Sigma}_{t\text{GP}}$ is the Hessian of the negative conditional log posterior density at $\tilde{\mathbf{m}}_{t\text{GP}}$:

$$\boldsymbol{\Sigma}_{t\text{GP}} = -\nabla^2 \log p_{t\text{GP}}(\mathbf{f}|\mathbf{y})|_{\mathbf{f}=\tilde{\mathbf{m}}_{t\text{GP}}} = \mathbf{K}_f^{-1} + \mathbf{W}_{t\text{GP}},$$

and

$$\mathbf{W}_{t\text{GP}} = -\nabla^2 \log p_{t\text{GP}}(\mathbf{y}|\mathbf{f})|_{\mathbf{f}=\tilde{\mathbf{m}}_{t\text{GP}}}, \quad W_{ii} = -(\nu+1)\frac{(y_i - f_i)^2 - \nu\sigma_\epsilon^2}{((y_i - f_i)^2 + \nu\sigma_\epsilon^2)^2}$$

is diagonal, since the likelihood factorizes over observations. The log likelihood for the approximate $q(\cdot)$ is

$$\log q(\mathbf{y}|\boldsymbol{\theta}) = \log p(\mathbf{y}|\mathbf{f}) - \frac{1}{2}\tilde{\mathbf{m}}_{t\text{GP}}^\top \mathbf{K}^{-1}\tilde{\mathbf{m}}_{t\text{GP}} - \frac{1}{2}\log |\mathbf{K}| - \frac{1}{2}\log |\mathbf{K}^{-1} + \mathbf{W}_{t\text{GP}}|.$$

Using (3.15), the approximate posterior distribution is also Gaussian $f(\mathbf{X}_*)|\mathcal{D} \sim \mathcal{N}(m_{t\text{GP}}(\mathbf{X}_*), s_{t\text{GP}}^2(\mathbf{X}_*))$, defined by its mean $m_{t\text{GP}}(\mathbf{X}_*)$ and covariance $s_{t\text{GP}}(\mathbf{X}_*)$:

$$
\begin{cases}
m_{t\text{GP}}(\mathbf{X}_*) = K(\mathbf{X}_*, \mathbf{X})\mathbf{K}_f^{-1}\tilde{\mathbf{m}}_{t\text{GP}}, \\
s_{t\text{GP}}(\mathbf{X}_*) = K(\mathbf{X}_*, \mathbf{X}_*) - K(\mathbf{X}_*, \mathbf{X})[\mathbf{K}_f + \mathbf{W}_{t\text{GP}}^{-1}]^{-1} K(\mathbf{X}, \mathbf{X}_*).
\end{cases}
\tag{3.16}
$$

Note the similarity of (3.16) to original GP equations (1.8)–(1.9) in Chap. 1: with Student-t likelihood the mode $\tilde{\mathbf{m}}_{t\text{GP}}$ plays the role of the adjusted observations $\mathbf{y}$ and $\mathbf{W}_{t\text{GP}}^{-1}$ replaces the noise matrix $\Sigma_\epsilon = \sigma_\epsilon^2 \mathbf{I}$. Critically, the latter implies that unlike (1.9) the posterior variance $s_{t\text{GP}}(\mathbf{x}_*, \mathbf{x}_*')$ is a function of both designs $\mathbf{X}$ and observations $\mathbf{y}$.

3.2.2 Student-t Process Regression with Student-t Noise

Instead of just adding Student-t likelihood to the observations, [149] proposed t-processes (TPs) as an alternative to GPs, deriving closed-form expressions for the marginal likelihood and posterior distribution of the t-process by imposing an inverse Wishart process prior over the covariance matrix of a GP model. TPs retain most of the appealing properties of GPs and can be more robust to model misspecification in very small samples $N \ll 100$. However, dealing with noisy observations is less straightforward with TPs, since the sum of two independent Student-t distributions has no closed form. One solution [149] is to incorporate the noise directly in the kernel, so that the data-generating mechanism is multivariate-t $\mathbf{y} \sim \mathcal{T}(v, f(\mathbf{X}), \mathbf{K}_f + \sigma_\epsilon^2 \mathbf{I})$, where the degrees of freedom are $v \in (2, \infty)$. The posterior predictive distribution is then $f(\mathbf{X}_*)|\mathcal{D} \sim \mathcal{T}(v + N, m_{\text{TP}}(\mathbf{X}_*), s_{\text{TP}}^2(\mathbf{X}_*))$, where [149]

$$
\begin{cases}
\widehat{m}_{\text{TP}}(\mathbf{X}_*) = K(\mathbf{x}_*, \mathbf{X})[\mathbf{K}_f + \sigma_\epsilon^2 \mathbf{I}]^{-1}\mathbf{y}, \\
s_{\text{TP}}^2(\mathbf{X}_*, \mathbf{X}_*^\top) = \dfrac{v + \hat{\eta} - 2}{v + N - 2}\left\{ K(\mathbf{X}_*, \mathbf{X}_*) - K(\mathbf{X}_*, \mathbf{X})[\mathbf{K}_f + \sigma_\epsilon^2 \mathbf{I}]^{-1} K(\mathbf{X}, \mathbf{X}_*)^\top \right\},
\end{cases}
$$

with $\hat{\eta} \overset{\triangle}{=} \mathbf{y}^\top [\mathbf{K}_f + \sigma_\epsilon^2 \mathbf{I}]^{-1}\mathbf{y}$.

Comparing with the regular GP, TP has the same posterior mean $m_{\text{TP}}(\mathbf{x}_*) = m(\mathbf{x}_*)$, but the posterior variance s_{TP}^2 depends on observations $\mathbf{y}$ and is inflated: $s_{\text{TP}}^2(\mathbf{x}_*) = \frac{v + \hat{\eta} - 2}{v + N - 2}s^2(\mathbf{x}_*)$. Moreover, the latent TP f and the noise are uncorrelated but not independent. Assuming the same hyperparameters, as N goes to infinity (practically speaking when $N \gg 40$), the above predictive distribution becomes Gaussian. Inference of TPs can be performed similarly as for a GP, for instance based on the marginal likelihood:

$$L_{\mathrm{TP}}(\theta|\mathbf{y}) = \frac{\Gamma(\frac{\nu+N}{2})}{((\nu-2)\pi)^{\frac{N}{2}}\Gamma(\frac{\nu}{2})}|\mathbf{K}_f + \sigma_\epsilon^2\mathbf{I}|^{-1/2} \times \left(1 + \frac{\mathbf{y}^\top[\mathbf{K}_f + \sigma_\epsilon^2\mathbf{I}]^{-1}\mathbf{y}}{\nu-2}\right)^{-\frac{\nu+N}{2}}.$$

(3.17)

One issue is estimation of the degrees of freedom ν, which plays a central role in the TP predictions. It is found in [149] that restricting ν to be small when maximizing (3.17) is important in order to avoid degenerating to the plain GP setup.

3.2.3 Gaussian Process GLM

A natural related setting are Generalized Linear Models (GLM) that replace the additive observation model in (1.5) with a different link function. This allows to consider for example discrete count data, where a natural choice is a Poisson observation likelihood, i.e. using a GP to model the latent mean output at input $\mathbf{x}$. Generally speaking, one specifies the distribution $p(y(\mathbf{x})|f(\mathbf{x}))$, where $f \sim \mathcal{GP}(\mu, k)$ is embedded in the link function of the GLM. Common cases include

$$y(\mathbf{x})|f(\mathbf{x}) \sim \mathrm{Poisson}(e^{-f(\mathbf{x})})$$

$$y(\mathbf{x})|f(\mathbf{x}) \sim \mathrm{Binomial}\left(N(\mathbf{x}), \phi(f(\mathbf{x}))\right),$$

where $\phi : \mathbb{R} \to [0, 1]$ is the sigmoid, probit, etc., transformation. Conditional on $\mathbf{f}$, the observations $\mathbf{y}$ are assumed independent like in the Gaussian case, meaning $p(\mathbf{y}|\mathbf{f})$ is a product and can be factored in closed form like in Eq. (3.14). The posterior $p(\mathbf{f}|\mathbf{y})$ is again intractable, requiring similar approximations like the LP method or expectation propagation (EP) [140, Chapter 3] approaches.

3.3 Multi-Output GPs

One may generalize GPs to consider vector-valued functions. This corresponds to jointly modeling a collection $\mathbf{y} = (y_1, \dots, y_L)$ of L outputs, all viewed as noisy maps $y_\ell(\mathbf{x}) = f_\ell(\mathbf{x}) + \epsilon_\ell(\mathbf{x})$ for each $\ell = 1, \dots, L$. The idea of Multi-Output GPs (MOGP) is to introduce dependence across outputs $\ell = 1, \dots, L$, so that knowledge of f_i provides information about f_ℓ and vice versa. This opens the door for information fusion, borrowing from an observation y_i to better learn a different f_ℓ, improving both hyperparameter fit and predictive accuracy.

Shared Covariance Structure Assume that each individual f_ℓ, $1 \le \ell \le L$, follows a GP $f_\ell \sim \mathcal{GP}(\mu_\ell, k_\ell)$. Encode inputs to their respective output by defining $\check{\mathbf{x}}_\ell \overset{\triangle}{=} (\mathbf{x}, \ell) \in \mathbb{R}^d \times \{1, \dots, L\}$ where the second coordinate is an output mapping.

The goal is to define a proper mean and covariance function so that we can apply all previous results to a GP f on the enlarged domain $\mathbb{R}^d \times \{1, \ldots, L\}$. In this case, the vector valued GP in $\mathbb{R}^L$ over $\mathbf{x} \in \mathbb{R}^d$ is

$$\mathbf{f}(\mathbf{x}) = [f(\check{\mathbf{x}}_{i,1}), \ldots, f(\check{\mathbf{x}}_{i,L})]^\top = [f_1(\mathbf{x}), \ldots, f_L(\mathbf{x})]^\top.$$

It is straightforward to define the corresponding mean function output-wise as $\mu(\check{\mathbf{x}}_\ell) = \mu_\ell(\mathbf{x})$. For the covariance, one requires $k(\check{\mathbf{x}}_\ell, \check{\mathbf{x}}'_{\ell'}) = k_\ell(\mathbf{x}, \mathbf{x}')$ when $\ell = \ell'$, meaning the resulting $\mathbf{K} \in \mathbb{R}^{LN \times LN}$ has L block-diagonals of size $N \times N$, one for each output.

A common way to proceed for the off-diagonal entries ($\ell \neq \ell'$) is to assume that the $\mathbf{x}$-kernel (denoted $\tilde{k}$) is the same across outputs. Then, sharing information across outputs is done through the $L \times L$ symmetric correlation matrix $\mathbf{B}$ with entries $r_{\ell,\ell'} \in [-1, 1]$ (and diagonal $r_{\ell,\ell} = 1$ for all ℓ). The covariance kernel $k : (\mathbb{R}^d \times \{1, \ldots, L\}) \times (\mathbb{R}^d \times \{1, \ldots, L\}) \to \mathbb{R}$ of f and its respective $LN \times LN$ covariance matrix $\mathbf{K}$ over $\mathcal{D} \times \{1, \ldots, L\}$ are

$$k(\check{\mathbf{x}}_\ell, \check{\mathbf{x}}'_{\ell'}) \stackrel{\triangle}{=} r_{\ell,\ell'} \tilde{k}(\mathbf{x}, \mathbf{x}'), \qquad \mathbf{K} \stackrel{\triangle}{=} \mathbf{B} \otimes \tilde{\mathbf{K}},$$

where $\otimes$ is the Kronecker product. This defines a valid kernel since the Kronecker product respects positive-definiteness. Note that the covariance matrix consists of L^2 blocks, each of size $N \times N$, where the (ℓ, ℓ')th block is $r_{\ell,\ell'} \cdot \tilde{\mathbf{K}}$. It is straightforward to confirm that $\text{corr}(f(\check{\mathbf{x}}_\ell), f(\check{\mathbf{x}}'_{\ell'})) = \text{corr}(f_\ell(\mathbf{x}), f_{\ell'}(\mathbf{x})) = r_{\ell,\ell'}$.

Along a similar line of reasoning, one can consider correlated noises, so that $\text{cov}(\epsilon(\check{\mathbf{x}}_{i,\ell}), \epsilon(\check{\mathbf{x}}_{j,\ell'})) = r^\epsilon_{\ell,\ell'} \sigma^2(\mathbf{x}_i) 1_{\{i=j\}}$ where $r^\epsilon_{\ell,\ell'} \in [-1, 1]$.

Intrinsic Coregionalization Model (ICM) Directly specifying the cross-covariance matrix $\mathbf{B}$ between all output pairs becomes impractical for large L [86]. A popular approach to address this is ICM [5, 23, 86], which models f over $\mathbb{R}^d \times \{1, \ldots, L\}$ by expressing each output f_ℓ as a linear combination of Q independent latent Gaussian processes u_q:

$$f(\check{\mathbf{x}}_\ell) = f_\ell(\mathbf{x}) = \sum_{q=1}^{Q} a_{\ell,q} u_q(\mathbf{x}), \quad \ell = 1, \ldots, L, \tag{3.18}$$

where the $a_{\ell,q}$ are coefficients forming an $L \times Q$ matrix A with ℓ, q entry $a_{\ell,q}$. The latent processes u_q are independent GPs with shared kernel $\tilde{k}(\mathbf{x}, \mathbf{x}')$. Under this model, the outputs are correlated through the shared latent processes. The covariance across inputs and outputs is

$$k(\check{\mathbf{x}}_\ell, \check{\mathbf{x}}'_{\ell'}) = \sum_{q=1}^{Q} \sum_{q=1}^{Q} a_{\ell,q} a_{\ell',q} \tilde{k}(\mathbf{x}, \mathbf{x}') = b_{\ell,\ell'} \tilde{k}(\mathbf{x}, \mathbf{x}'), \tag{3.19}$$

where $b_{\ell,\ell'} = a_\ell^\top a_{\ell'}$ are the entries of the $L \times L$ *coregionalization matrix* $\mathbf{B} = AA^\top$. The difference from before is that this matrix $\mathbf{B}$ has rank at most Q. The full covariance matrix of the vector $\mathbf{f} = (\mathbf{f}_1^\top, \ldots, \mathbf{f}_L^\top)^\top \in \mathbb{R}^{LN}$ is again

$$K = \mathbf{B} \otimes \tilde{K}. \tag{3.20}$$

By choosing $Q \ll L$, the number of hyperparameters in $\mathbf{B}$ is reduced from $\frac{L(L+1)}{2}$ to $L \times Q$, as now the model is parameterized through the $a_{\ell,q}$ instead of $r_{\ell,\ell'}$ across all pairs. This procedure is advantageous for large L, analogous to techniques like principal component analysis (PCA) or singular value decomposition (SVD). The rank Q can be selected by the user or determined via cross-validation. A popular generalization of the ICM is the Linear Model of Coregionalization [6, 102, 106] which uses R collections of latent processes with varying kernels $k_r(\cdot, \cdot)$ and ranks $Q_r, r = 1, \ldots, R$.

3.4 Localization

Multiple approaches exist to localize the GP predictive equations (1.8)–(1.9). Localization is helpful either to address large datasets, where regular GP suffers from complexity of order $O(N^3)$, or alternatively to address *non-stationarity*, i.e. different data features (such as roughness or correlation structure) across different regions of the input space. For example, for the option pricing application in Chap. 4, the response has a strong curvature (hence relatively small lengthscale) at-the-money, and is nearly linear deep out-of-the-money (where option value is zero) or deep in-the-money. A global fit might then lead to underestimating the lengthscale, generating spurious oscillations away from at-the-money as the kernel (and especially posterior samples) refuses to be flat/linear. In this section we provide a brief survey of the main localization or divide-and-conquer techniques.

Partitioning Partitioning means fitting several different GPs on different sub-domains of the input space X and is effectively equivalent to making the covariance matrix $\mathbf{K}$ block-diagonal. The blocks can be imposed a priori or, in advanced implementations, chosen adaptively. While divide-and-conquer is guaranteed to save computational effort, haphazard partitioning might severely affect accuracy. Indeed, a data-driven partitioning allows to simultaneously achieve computational gains through smaller matrices to decompose/invert and to capture disparate dependence structures, killing two birds with one stone. The major trade-off is how to "glue" the partitions, which ultimately affects the continuity of the fit across partition boundaries. We refer to Gramacy and Lee [81] for a tree-based dynamic partitioning approach and to Kim et al. [96] for a Voronoi tesselation based partitioning. The respective implementations are nontrivial and we caution the readers to rely on existing packages, such as tgp [77]. For the tree based approaches, there are many tuning choices, such as rules for splitting and pruning trees and the interaction between the hyperparameters of adjoining partitions. Dynamic trees is a further extension that facilitates streaming data, where the GP is sequentially updated as new data is assimilated.

Nearest Neighbor GPs As we saw in (1.10), the GP prediction $\mu(\mathbf{x}_*)$ can be understood as a weighted average of the input data $\mathbf{y}$. For a large dataset, many of the corresponding weights, namely those from far-away inputs $\mathbf{x}$'s will be close to zero. This suggests an approximation method that replaces the original GP prediction with a smaller average, over just the "neighbors" of $\mathbf{x}_*$. Conceptually, this corresponds to approximating $p(f(\mathbf{x}_*)|\mathbf{y}) \simeq p(f(\mathbf{x}_*)|\mathbf{y}_{N(\mathbf{x}_*)})$ for a properly defined neighborhood map $N : \mathcal{X} \rightarrow 2^{\mathbf{X}}$ mapping inputs into subsets of the training set. A canonical such mapping is the n-nearest neighbors: $N(\mathbf{x}_*) = \{\mathbf{x} \in \mathbf{X} : \|\mathbf{x}_* - \mathbf{x}\|$ is among the n-nearest $\}$ for some metric $\| \cdot \|$ (e.g. the Euclidean one on $\mathbb{R}^d$) [39]. The corresponding posterior NNGP equations are same as the main GP formulas (1.8), substituting throughout the full $K(\cdot, \mathbf{X})$ with the smaller $K(\cdot, N(\mathbf{x}_*))$. Compared to the classical GP, NNGP constructs a sparse precision matrix, zeroing out all terms of the covariance structure where the inputs are not neighboring. The related *Vecchia approximation* [92, 93] is another way to make $\mathbf{K}^{-1}$ sparse.

Local GPs A simplification of the above construction is to altogether ignore all non-neighboring data. Such local GPs [80] dispense with global structure and focus on prediction only. This means that like in local regression (LOESS), a separate $\mathbf{K}^{\mathbf{x}_*}$ matrix is generated on the fly for a given $\mathbf{x}_*$. The idea is that this can keep $\mathbf{K}^{\mathbf{x}_*}$ very small and very flexible. Think building a matrix of size 50×50 even when there are thousands of inputs. The best analogy is again of a neighborhood—to predict the response at $\mathbf{x}_*$ one utilizes the training outputs at the *neighbors* of $\mathbf{x}_*$, generally implemented through a nearest-neighbor search. For GP purposes, this subdesign is generally augmented with a few faraway inputs $\mathbf{x}_i$ in order to capture some global dependence structure. Local GPs are best suited for massive datasets with a limited number of needed predictions; they should not be used for large-scale predictive tasks as a nontrivial amount of overhead is expended to build each new $\mathbf{K}^{\mathbf{x}_*}$. To better approximate the local structure, local GPs refit hyperparameters for each $\mathbf{x}_*$, which adds flexibility to capture behavior that is nonstationary globally, but also leads to overfitting and globally non-smooth predictive surfaces. In contrast, NNGP maintains a single set of hyperparameters and just focuses on making $\mathbf{K}$ and $\mathbf{K}^{-1}$ sparse. A further perspective is to think of local GPs as soft partitioning. Instead of creating hard boundaries in the input space and fitting a separate GP in each region, a local GP builds a continuum of GPs indexed by the predictive location—in analogue to moving from regular multivariate regression to a local (LOESS-style) generalization.

3.4.1 Inducing Points

A popular localization strategy involves *inducing points*, as introduced in [153]. The core idea of inducing points is to overcome the computational burden that is cubic in the number of data points by selecting a few points in the input space to treat as "landmarks" for prediction. To this end, consider a collection $\bar{\mathbf{X}} \triangleq \{\bar{\mathbf{x}}_i\}_{i=1}^{m}$ of $m \ll N$ inducing points with corresponding latent function values $\bar{f}_m \triangleq f(\bar{\mathbf{x}}_m)$. These inducing points are intended to fully inform new predictions, adding an independence assumption between f_* and $\mathbf{y}$ given $\bar{\mathbf{f}}$. Hence the predictive GP f_* at a new location $\mathbf{x}_*$ is factored as

$$p(f_*|\mathbf{y}, \bar{\mathbf{f}}) = p(f_*|\bar{\mathbf{f}})p(\bar{\mathbf{f}}|\mathbf{y})$$

An alternative interpretation is to construct a rank-m approximation $\tilde{\mathbf{K}}_m$ to the full covariance matrix $\mathbf{K}$, namely through the approximation

$$\mathbf{K} \simeq \tilde{\mathbf{K}}_m \triangleq K(\mathbf{X}, \bar{\mathbf{X}})K(\bar{\mathbf{X}}, \bar{\mathbf{X}})^{-1}K(\bar{\mathbf{X}}, \mathbf{X}) \tag{3.21}$$

and modify the prior to be

$$p(\mathbf{f}) = \mathcal{N}(\mathbf{0}, \tilde{\mathbf{K}}_m + \mathbf{\Lambda}),$$

where the diagonal matrix $\mathbf{\Lambda} = \text{diag}(\mathbf{K} - \tilde{\mathbf{K}}_m)$ has entries $\lambda_i \triangleq k(\mathbf{x}_i, \mathbf{x}_i) - K(\mathbf{x}_i, \bar{\mathbf{X}})K(\bar{\mathbf{X}}, \bar{\mathbf{X}})^{-1}K(\mathbf{x}_i, \bar{\mathbf{X}})^\top$. This Fully Independent Training Conditional (FITC) approach means that the marginal log-likelihood is adjusted to

$$\ell_{FITC} = -\frac{N}{2}\log 2\pi - \frac{1}{2}\log|\tilde{\mathbf{K}}_m + \mathbf{\Lambda} + \sigma_\epsilon^2\mathbf{I}| - \frac{1}{2}\mathbf{y}^\top(\tilde{\mathbf{K}}_m + \mathbf{\Lambda} + \sigma_\epsilon^2\mathbf{I})\mathbf{y}. \tag{3.22}$$

Using the same GP prior $\bar{\mathbf{f}} \sim \mathcal{MVN}(0, K(\bar{\mathbf{X}}, \bar{\mathbf{X}}))$ and the posterior $p(\mathbf{y}|\bar{\mathbf{f}}) = \mathcal{MVN}(K(\mathbf{X}, \bar{\mathbf{X}})K(\bar{\mathbf{X}}, \bar{\mathbf{X}})^{-1}\bar{\mathbf{f}}, \mathbf{\Lambda} + \sigma_\epsilon^2\mathbf{I})$ we can then integrate out over $\bar{\mathbf{f}}$ giving the Gaussian posterior at predictive locations $\mathbf{X}_*$ as $f_*(\mathbf{X}_*) \sim \mathcal{N}(m_{IP}(\mathbf{X}_*), s_{IP}(\mathbf{X}_*))$ with

$$m_{IP}(\mathbf{X}_*) \triangleq K(\mathbf{X}_*, \bar{\mathbf{X}})\mathbf{Q}_m^{-1}K(\bar{\mathbf{X}}, \mathbf{X})(\mathbf{\Lambda} + \sigma_\epsilon^2\mathbf{I}_N)^{-1}\mathbf{y} \tag{3.23}$$

$$s_{IP}(\mathbf{X}_*) \triangleq K(\mathbf{X}_*, \mathbf{X}_*) - K(\mathbf{X}_*, \bar{\mathbf{X}})(K(\bar{\mathbf{X}}, \bar{\mathbf{X}})^{-1} - \mathbf{Q}_m^{-1})K(\bar{\mathbf{X}}, \mathbf{X}_*) \tag{3.24}$$

$$\mathbf{Q}_m \triangleq K(\bar{\mathbf{X}}, \bar{\mathbf{X}}) + K(\bar{\mathbf{X}}, \mathbf{X})(\mathbf{\Lambda} + \sigma^2\mathbf{I}_n)^{-1}K(\mathbf{X}, \bar{\mathbf{X}}). \tag{3.25}$$

The key computational saving is that the matrix $\Lambda + \sigma_\epsilon^2 \mathbf{I}_N$ is of size N but is now diagonal, so fast to invert, while the dense matrix $\mathbf{Q}_m$ is of size $m \times m$. Hence, the two expensive operations in (3.23) is the matrix-matrix multiplication $O(N^2 m)$ and the matrix inversion $O(m^3)$ in (3.24), reducing to quadratic cost in N when $m \ll N$.

The quality of the inducing point approximation depends on the choice of the inducing points $\bar{\mathbf{X}}$. A simple solution is to make $\bar{\mathbf{X}}$ to be a (random) subset of the training $\mathbf{X}$, which leads to the Nystrom approximation. Alternatively, $\bar{\mathbf{X}}$ can be directly picked by the user, but generally speaking they should be optimized as part of fitting the GP. Snelson and Ghahramani [153] treat them as kernel hyperparameters, effectively parametrizing the log-likelihood (3.22) by $\bar{\mathbf{X}}$.

Variational Gaussian Processes

Variational methods take the inducing point approach further. Similar logic to above shows

$$p(f_*|\mathbf{y}) = \int p(f_*|\bar{\mathbf{f}})p(\bar{\mathbf{f}}|\mathbf{y})d\bar{\mathbf{f}}. \tag{3.26}$$

It is unrealistic to find inducing points $\bar{\mathbf{X}}$ so that Eq. (3.26) holds exactly, meaning an approximation must be used somewhere. Titsias [160] cleverly posited the approximation $\phi(\bar{\mathbf{f}}) \approx p(\bar{\mathbf{f}}|\mathbf{y})$, where under ϕ, $\bar{\mathbf{f}} \sim N(\mu, \mathbf{A})$ is a "free" variational Gaussian distribution. Replacing $p(\bar{\mathbf{f}}|\mathbf{y})$ with $\phi(\bar{\mathbf{f}})$ in Eq. (3.26) shows the same reduction in computational cost to the traditional inducing point approach with the added flexibility of μ and $\mathbf{A}$. To determine the inducing points and $\mu, \mathbf{A}$, the variational approach minimizes the KL divergence between $\phi(\bar{\mathbf{f}})$ and $p(\bar{\mathbf{f}}|\mathbf{y})$. This can be analytically simplified to maximizing [12]

$$\ell_{VGP} = -\frac{N}{2}\log 2\pi - \frac{1}{2}\log|\tilde{\mathbf{K}}_m + \sigma_\epsilon^2 \mathbf{I}| - \frac{1}{2}\mathbf{y}^\top(\tilde{\mathbf{K}}_m + \sigma_\epsilon^2 \mathbf{I})\mathbf{y} - \frac{1}{2\sigma_\epsilon^2}\mathrm{tr}(\mathbf{K} - \tilde{\mathbf{K}}_m) \tag{3.27}$$

where $\tilde{\mathbf{K}}_m = \mathrm{cov}(\mathbf{f}|\bar{\mathbf{f}})$ from (3.21) can be thought of a regularizer on the original inducing point problem. There exists analytical maximizers of μ and $\mathbf{A}$:

$$\mu = \frac{1}{\sigma_\epsilon^2}K(\bar{\mathbf{X}}, \bar{\mathbf{X}})\left[K(\bar{\mathbf{X}}, \bar{\mathbf{X}}) + \sigma_\epsilon^{-2}K(\bar{\mathbf{X}}, \mathbf{X})K(\mathbf{X}, \bar{\mathbf{X}})\right]^{-1}K(\bar{\mathbf{X}}, \mathbf{X})\mathbf{y},$$

$$\mathbf{A} = K(\bar{\mathbf{X}}, \bar{\mathbf{X}})\left[K(\bar{\mathbf{X}}, \bar{\mathbf{X}}) + \sigma_\epsilon^{-2}K(\bar{\mathbf{X}}, \mathbf{X})K(\mathbf{X}, \bar{\mathbf{X}})\right]^{-1}K(\bar{\mathbf{X}}, \bar{\mathbf{X}})$$

so that objective in optimizing (3.27) is to choose $\bar{\mathbf{X}}$ (and optionally m). Replacing $p(\bar{\mathbf{f}}|\mathbf{y})$ with $\phi(\bar{\mathbf{f}})$ in (3.26) gives approximate posterior GP mean

$$m_{VGP}(\mathbf{x}_*) = K(\mathbf{x}_*, \bar{\mathbf{X}}) K(\bar{\mathbf{X}}, \bar{\mathbf{X}})^{-1} \mu,$$

$$= \frac{1}{\sigma_\epsilon^2} K(\mathbf{x}_*, \bar{\mathbf{X}}) \left[K(\bar{\mathbf{X}}, \bar{\mathbf{X}}) + \sigma_\epsilon^{-2} K(\bar{\mathbf{X}}, \mathbf{X}) K(\mathbf{X}, \bar{\mathbf{X}}) \right]^{-1} K(\bar{\mathbf{X}}, \mathbf{X}) \mathbf{y}$$

and a similar (longer) expression for $s_{VGP}(\mathbf{x}_*, \mathbf{x}'_*)$. While the posterior equations are very similar between VGP and FITC, the optimization problems lead to quite different fits; recent literature generally recommends VGP over FITC even though it is more computationally costly at the MLE stage.

3.5 Updating Equations for GPs

Often in application one considers *streaming* data, where the dataset $\mathcal{D}$ is itself changing over the course of the modeling. The most common situation is when $\mathcal{D}$ *grows* over time, i.e. new data samples arrive. To emphasize the changing size of the training set, we explicitly index $\mathcal{D}$, as well as the GP posterior mean and variance, by $n = |\mathcal{D}_n|$. In the most common situation, we assume that starting with $\mathcal{D}_n$ one more data point is added, i.e., we augment $(\mathbf{x}_1, y_1), \ldots, (\mathbf{x}_n, y_n)$ with $(\mathbf{x}_{n+1}, y_{n+1})$. The multivariate Gaussian conditioning equations facilitate an efficient *update* of the GP posterior equations. Thus, rather than compute the new $m_{n+1}(\cdot)$ from scratch, one may update it from the earlier $m_n(\cdot)$ and similarly for updating the posterior standard deviation $s_{n+1}(\cdot)$ based on $s_n(\cdot)$. Beyond a conceptual appeal, updating is also computationally efficient. The main expense in GP prediction is to invert the large covariance matrix $\mathbf{K}_{n+1}$, a step replaced below with reusing $\mathbf{K}_n^{-1}$ and a few matrix-vector computations.

The updating is achieved through the Woodbury identities (A.2). Let $\mathbf{X}_n = [\mathbf{x}_1, \ldots, \mathbf{x}_n]$. We present the rank-1 update [79] based on the block decomposition:

$$\mathbf{K}_{n+1} = \begin{bmatrix} \mathbf{K}_n & K(\mathbf{X}_n, \mathbf{x}_{n+1}) \\ K(\mathbf{x}_{n+1}, \mathbf{X}_n) & k(\mathbf{x}_{n+1}, \mathbf{x}_{n+1}) \end{bmatrix} \quad \Rightarrow \quad \mathbf{K}_{n+1}^{-1} = \begin{bmatrix} \mathbf{K}_n^{-1} + \mathbf{U}_n & \mathbf{g}_n \\ \mathbf{g}_n^\top & v_n^{-1} \end{bmatrix}, \tag{3.28}$$

where the scalar $v_n = s_n^2(\mathbf{x}_{n+1}) = k(\mathbf{x}_{n+1}, \mathbf{x}_{n+1}) - K(\mathbf{x}_{n+1}, \mathbf{X}_n)\mathbf{K}_n^{-1} K(\mathbf{X}_n, \mathbf{x}_{n+1})$ is the step-n GP posterior variance (1.9) at the input $\mathbf{x}_{n+1}$, and the other two terms are the $n \times 1$ vector $\mathbf{g}_n \stackrel{\triangle}{=} -v_n^{-1}\mathbf{K}_n^{-1} K(\mathbf{X}_n, \mathbf{x}_{n+1})$ and the $n \times n$ matrix $\mathbf{U}_n = \mathbf{g}_n \mathbf{g}_n^\top v_n$. Observe that we obtain the inverse $\mathbf{K}_{n+1}^{-1}$ through the pre-computed $\mathbf{K}_n^{-1}$ plus several vector operations that take a total time of $O(n^2)$. By evaluating the posterior variance formula (1.9) with (3.28), one obtains

$$s_{n+1}^2(\mathbf{x}_*) = s_n^2(\mathbf{x}_*) - K(\mathbf{x}_*, \mathbf{X}_n)\mathbf{g}_n \mathbf{g}_n^\top s_n^2(\mathbf{x}_{n+1}) K(\mathbf{X}_n, \mathbf{x}_*)$$

$$- 2K(\mathbf{x}_*, \mathbf{X}_n)\mathbf{g}_n^\top k(\mathbf{x}_{n+1}, \mathbf{x}_*) + k(\mathbf{x}_{n+1}, \mathbf{x}_*)^2 s_n^2(\mathbf{x}_{n+1}) \tag{3.29}$$

which provides updating for the posterior variance; covariance follows similarly. These equations can be straightforwardly generalized to a block version to allow batch updates. Analogously,

$$\log |\mathbf{K}_{n+1}| = \log |\mathbf{K}_n| + \log s_n^2(\mathbf{x}_{n+1}) \tag{3.30}$$

so that the determinant of the covariance matrix changes by the variance of the new input $\mathbf{x}_{n+1}$ added to the existing experimental design $\mathbf{X}_n$.

Well-Conditioned Training Designs
The expression (3.30) provides insight into constructing numerically stable experimental designs for GPs. It indicates that relatively ill-conditioned matrices whose determinants are small compared to their size arise due to adding data at low-variance locations. Therefore, to obtain well-conditioned $\mathbf{K}_n$ (as well as a better optimization landscape for hyperparameter optimization) one should augment with data at high-variance regions, enforcing the concept of exploration.

Further Reading

With the GP universe in the midst of active development, there are many research strands that explore extensions and modifications of the core concepts. This chapter is therefore just scratching the surface, with the main take-away to outline some of the major themes, such as other observation likelihoods, computational approaches to speed up the evaluation of the GP predictive equations, and the generalization to multiple outputs. The second half of Gramacy [79] is one starting point for more reading. There are numerous further extensions to approximate large-scale GPs, such as Ensemble GPs [58], KISS-GP [169], etc.; a recent survey, which contains an extensive bibliography, on scaling GPs to many thousands of inputs is [105]. Multi-output modeling is another rapidly growing area [6]. Beyond the presented MOGP approach, we mention the co-kriging hierarchical setup [135] linking to multi-fidelity models and multi-level schemes.

The various extensions also motivate ongoing development of additional GP software libraries, see for instance the Python packages MOGPTK [42] which emphasizes multi-output GPs, and GPyTorch [65] which are based on efficient linear algebra approximation routines to significantly increase computational efficiency.

Chapter 4
Option Pricing and Sensitivities

We pivot to describing applications of the GP methodology in quantitative finance. In this Chapter we focus on employing GPs as pricing *surrogates*, providing a fast prediction of contracts' values as a function of their parameters. Thus, the "input–output" relationship in previous chapters translates into mapping a state $\mathbf{x}$ (which might include underlying prices, but also deterministic contract parameters like its strike) into a option price $P(\mathbf{x})$.

In what follows, we assume a probability space with filtration $\mathbb{F} = (\mathcal{F}_t)_{t \geq 0}$ indexed by t. Usually this is the natural filtration generated by some state process $(\mathbf{X}_t)_{t \geq 0}$, where $\mathbf{X}_t$ is chosen so that relevant processes are Markovian when needed. Financially, option prices are obtained from the risk-neutral pricing formula

$$P(\mathbf{x}) \triangleq \mathbb{E}^Q[\, \cdot \mid \mathbf{X}_t = \mathbf{x}], \tag{4.1}$$

and the main idea is to proxy the true $P(\cdot)$ with a GP prior $f \sim \mathcal{GP}(\mu, k)$, yielding point estimates $\hat{P}(\mathbf{x})$ via the posterior mean (equivalently, MAP) $m(\cdot)$ of $f_*(\cdot)$. Once a GP representation of the option prices has been learned, $\hat{P}(\cdot)$ replaces the traditional numerical pricing procedure.

4.1 Learning to Price Options

The financial sector's demand for up-to-the-minute recalculations of contract prices, hedging ratios, and risk exposures necessitates models that can adapt swiftly and accurately to market changes. Direct methods, such as solving a partial differential equation to obtain $P(\mathbf{x})$, while reliable, struggle with the computational burden posed by complex models and large data volumes. This bottleneck motivates the emergence of statistical surrogates that enable the estimation of option prices from a set of inputs, such as market conditions and option characteristics. Such valuation surrogates start with a training repository, infer the input-output relationship and then use the surrogate predictions instead of the time-consuming existing alternative.

In this context, GPs offer a flexible and probabilistic framework that is organically aligned with the parametric nature of derivative valuation models, which conceptualize the pricing function as a mapping from inputs to outputs. By abstracting this mapping into a statistical learning problem, GPs facilitate a more efficient exploration of the price landscape compared to deterministic regression models. This efficiency stems from GPs' ability to act as both interpolators across a broad input domain and as flexible models that can adjust to new data without the need for retraining from scratch. Moreover, a probabilistic surrogate like GP provides error estimates which account for covariance of the derivative prices over the test points, as opposed to frequentist machine learning techniques, such as support vector machines or neural networks, which only provide point estimates. A high uncertainty in a prediction might result in a GP model estimate being rejected in favor of either retraining the surrogate or reverting to a different (slower) pricer. The GP uncertainty quantification also enables simulation based methods to direct the sampling to achieve certain margins of error.

To set the stage, consider learning a scalar output, such as a Call price, as a function of a scalar input, say maturity $T \mapsto P(T)$, given by the risk-neutral pricing formula

$$P(T) \stackrel{\triangle}{=} \mathbb{E}^Q \left[e^{-rT}(S_T - \mathcal{K})_+ | S_0 = s \right],$$

keeping the strike $\mathcal{K}$, the interest rate r and initial price of underlying s fixed. Rather than doing any mathematics, we take a statistical approach, aiming to agnostically infer the shape of $P(\cdot)$ from some training data. As a basic strategy, we can start with a grid of T-values, $\{T_1, \ldots, T_N\}$. For each T_n one runs the standard pricer (say a Monte Carlo simulator, see (4.4) below) to obtain the respective option price $P_n = P(T_n)$. Then for any arbitrary $T \in [T_1, T_N]$ in the range, one estimates the respective option price $\hat{P}(T)$ by linear interpolation of the training $\{P_n\}$'s. A second choice is to employ a regression model. If, for example, we use polynomials up to order 2,

$$\hat{P}(T) \stackrel{\triangle}{=} \beta_0 + \beta_1 T + \beta_2 T^2 \tag{4.2}$$

then estimated option prices are obtained by inferring the coefficients $\beta = (\beta_0, \beta_1, \beta_2)$, most commonly done by minimizing mean-squared errors,

$$\hat{\beta} = \arg \min_{\beta \in \mathbb{R}^3} \sum_{n=1}^{N} (\beta_0 + \beta_1 T_n + \beta_2 T_n^2 - P_n)^2.$$

The quadratic proxy (4.2) would tend to perform better than linear interpolation when the training dataset size N is moderate, but would still exhibit bias due to limited flexibility. Moving up to the next rung of complexity, a popular practitioner technique is to employ cubic *spline* interpolation. Splines allow for increased surrogate flexibility as N grows.

Finally, we are led to *non-parametric* approximators like the GP. The latter provides an expressive fit by dispensing with the parametric nature of the linear regression and moreover organically scales to multiple dimensions (unlike splines). As in Chap. 1, given a training set $\mathcal{D} = \{(P_n, T_n)\}_{n=1}^{N}$ we fit a GP model for the relation between inputs T_n and corresponding training prices P_n. Once trained, the GP surrogate is used to calculate option values for unobserved maturities T's. An important feature is that the GP learns the posteriors over the output space without knowing the functional form of the map between input and output or the data generation process. Thus, the underlying asset dynamics are kept separate from the GP regression; in fact the surrogate is agnostic to whether the training P_n's come from an option pricing model or real observations, and is therefore often referred to as "model-free" by practitioners. However, if the GP is trained based on outputs from a fixed framework (say the Heston model (4.3)) then of course its estimates tacitly depend on that framework.

To set the stage before diving into full generality, we first consider learning the Heston pricing formula as a function of maturity T (see [37, Section 3.3] for a similar case study). This is a 2-factor model driven by the bivariate correlated Wiener process $(W^{(1)}, W^{(2)})$. The price dynamics under Q are

$$dS_t = r S_t dt + \sigma \sqrt{v_t} S_t \, dW_t^{(1)}$$

with

$$dv_t = \kappa(\theta - v_t)dt + \eta\sqrt{v_t}\, dW_t^{(2)}, \qquad d\langle W_t^{(1)}, W_t^{(2)}\rangle = \rho dt. \qquad (4.3)$$

We fix the parameters of at-the-money Calls with $S_0 = 100, \mathcal{K} = 100, r = 0.05, v_0 = 0.09, \theta = 0.04, \eta = 0.5, \kappa = 1, \rho = -0.5$. Given the one-dimensional input T, our training output is the at-the-money Call price (computed via numerical quadrature of Laguerre–Gauss with degree 30). For the training data we select Call prices from 11 non-uniformly spaced maturities T_n, namely of 2, 3, 4, 5, 7, 9, 12, 15, 18, 21, 24 months. We then test over the range $T \in [1/12, 2]$ years. Figure 4.1 compares the resulting fits of the quadratic polynomial, a cubic spline with 4 knots and a GP with Matérn 5/2 covariance. All the methods are fitted by minimizing squared errors between training option prices and surrogate predictions. In the left panel, all methods match the general curvature of the ground truth; however the right panel reveals a systematic issue. The polynomial cannot properly capture the shape and ends up overfitting (a common feature for under-parametrized, unregularized regression). Both the cubic spline and the GP do much better, in fact the cubic spline provides a near-perfect fit. This is not surprising,

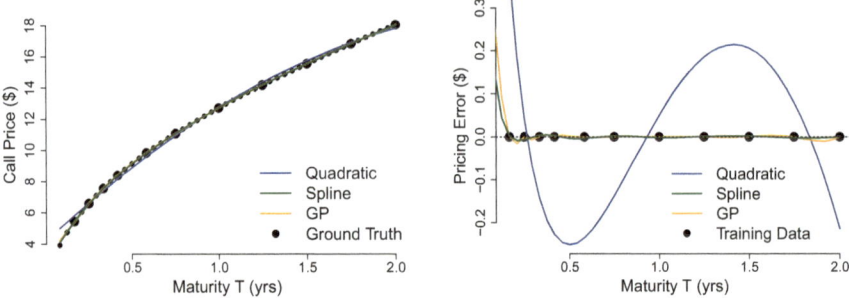

Fig. 4.1 Learning Call prices in the Heston Model. *Left:* $T \mapsto \hat{P}(T)$ for three different regressors. *Right:* pricing error $\hat{P}(T) - P(T)$ relative to the ground truth

but it is worth mentioning that there is no good multi-dimensional generalization of the cubic spline (two-dim. splines are common, but higher dimensions become cumbersome), while the GP is trivial to scale to higher dimensions. See De Spiegeleer et al. [41] for an extended case study of learning the Heston model; that reference demonstrates the speed-up of GPs relative to direct Monte Carlo methods and tolerable pricing accuracy, as well as the increased expressibility of GPs compared to cubic spline interpolation. Training a GP is more time-consuming than the simpler regressors, but it is done only once and hence not a material concern for deployment. This case study illustrates another point, namely that financial pricing functions are generally easy to statistically learn, as they tend to be smooth and well behaved across the input space. Thus, it comes not as a surprise that one is able to interpolate the entire Heston formula from less than a dozen (here noiseless) training samples.

The above logic carries over to more complex settings encountered on trading desks. De Spiegeleer et al. [41] present several such case studies using GPR, including parametrically learning American option prices, exotic option prices and hedge ratios, as well as de-Americanizing American option values. They also discuss learning a Gamma profile addressed later in this Chapter.

4.2 Surrogate Ingredients

Transitioning from the illustrative 1d Heston model example, we are now ready to outline a structured exposition of constructing GP surrogate models which can be applied in a setting of more general inputs and more complex pricing models.

Inputs and Outputs In order to construct GP surrogates for option pricing, one must specify the respective inputs $\mathbf{x}$ and outputs y. For $\mathbf{x}$ the pricing features may include not just the relevant stochastic states like underlying price, but also *option* parameters λ, and *model* parameters θ. For Calls and Puts, the most common features are asset price S, time t, strike $\mathcal{K}$ (an example of λ) and volatility σ (an example of θ). For many models, the option pricing function is homogeneous of degree one with respect to S and $\mathcal{K}$, making it common to reparametrize via their ratio known as the *moneyness* $M \overset{\triangle}{=} S/\mathcal{K}$. This enables to have a more stationary training set, reduces overfitting [64] and lowers the dimension of the inputs yielding computational efficiency.

Similarly, rather than separately looking at time t and option maturity T, it is common to re-parameterize the surrogate via time-to-maturity $\tau = T - t$. In the context of gradient boosting, [40] suggest using features, such as discount factors e^{-rT}, and first-order interactions between the parameters of the stochastic model (e.g. $\kappa \cdot \eta$ in the Heston model (4.3)). Given its prominence, the volatility σ is often used as a feature. In local or stochastic volatility models, one often includes the Black–Scholes implied volatility σ_{Imp}. At least conceptually, using separable GP kernels one may easily handle settings with 10+ features, which is needed to fully describe common financial models. For example, the simplest case of a Black-Scholes model for Call pricing already has 6 parameters $S_0, \mathcal{K}, T, r, q, \sigma$; De Spiegeleer et al. [41] present an example to learn the down-and-out barrier Put with barrier H in a Heston model with 10 pricing features $\mathbf{x} = [\mathcal{K}, H, T, r, q, \kappa, \rho, \theta, \eta, v_0]$. More complex products, like Bermudan swaptions in LIBOR models, have valuation functions with hundreds of inputs, involving all the properties of the underlying swap and option exercise schedule.

The most common surrogate output y is the option price P; if working with moneyness, the output could also be option price divided by its strike $P/\mathcal{K}$. The training prices y may come from a deterministic solver, like the numerical quadrature approximation for the Heston pricer mentioned above, or from a Monte Carlo simulation. In the latter case, we recall the classical sample average estimator of the European option with payoff functional $G(S_T)$ based on $\check{N}$ simulations:

$$y(\mathbf{x}) = \frac{1}{\check{N}} \sum_{n=1}^{\check{N}} e^{-r(T-t)} G(S_{T,n}(\mathbf{x})) \simeq \mathbb{E}^Q \left[e^{-r(T-t)} G(S_T) | \mathbf{x} \right], \qquad (4.4)$$

where $S_{T,n}(\mathbf{x})$ is the nth i.i.d. sample for a given initial state $\mathbf{x}$ under the pricing measure Q. The Monte Carlo training approach can be straightforwardly applied to path-dependent contracts where pricing is time consuming, such as barrier options or American options. The sampling variance of y in (4.4) depends on $\mathbf{x}$, for example being higher in-the-money than out-of-the-money where most simulated payoffs are zero. Hence, such training data is intrinsically heteroskedastic. Accordingly, it is important to record the empirical standard deviation of the $\check{N}$ payoffs to obtain the plug-in estimator $\hat{\sigma}(\mathbf{x})$ for the input-dependent noise variance parameter:

$$\hat{\sigma}^2(\mathbf{x}) = \frac{1}{\check{N}-1} \sum_{n=1}^{\check{N}} \left(y(\mathbf{x}) - e^{-r(T-t)} G(S_{T,n}(\mathbf{x})) \right)^2. \qquad (4.5)$$

In turn, the collection $\{(\mathbf{x}_i, y(\mathbf{x}), \hat{\sigma}(\mathbf{x}_i))\}_{i=1}^{N}$ can be used to construct a heteroskedastic GP, for example using the stochastic kriging approach, cf. Sect. 3.1.

Training Procedure We advance to the construction of the training dataset $\mathcal{D} = \{(\mathbf{x}_i, y_i)\}_{i=1}^{N}$ or $\{(\mathbf{x}_i, y_i, \hat{\sigma}(\mathbf{x}_i))\}_{i=1}^{N}$. As an illustrative example, consider revisiting the Heston model for a European call option in full detail. Here each $\mathbf{x}_i, i = 1, \ldots, N$ encompasses a combination of product, market, and model parameters: strike price ($\mathcal{K}$), maturity (T), interest rate (r), dividend yield (q) and the model-specific parameters, such as $\kappa, \rho, \theta, \eta, v_0$. To construct $\mathcal{D}$, one should first consider the potential parameter configurations $\bar{\mathcal{D}}$ of $(\mathbf{x}, \cdot) \in \mathcal{D}$. This step is pivotal since the surrogate aims to minimize estimation errors specifically within $\mathcal{D}$, meaning it must be representative of $\bar{\mathcal{D}}$, which in turn must be sufficient for the task at hand. Balancing between a broad $\bar{\mathcal{D}}$ for generality and a narrower one for reduced complexity and training efficiency is crucial [57]. For instance, training a model for a Bermudan swaption might involve fixing specific trade properties while varying market data inputs, leading to a more manageable training scope. Similarly, an appropriate sampling distribution for $\mathcal{D}$ is also critical. Although uniform sampling is a common default [41], tailored non-uniform distributions often yield better training outcomes. Intuitively, this approach is beneficial to avoid over-representation in low-interest areas, such as scenarios with zero option value, and in focusing on regions with significant valuation changes. As an example, [41] samples the volatility σ from a bounded exponential distribution to emphasize smaller volatilities, enhancing the model's performance in those areas.

For any given $\mathbf{x}$, the associated y are obtained as mentioned above. Here it is crucial to be thoughtful regarding noise, which can stem from observational uncertainties or the intrinsic randomness of Monte Carlo simulations (which clearly depends on $\mathbf{x}$). Even in the noiseless case, it is helpful to introduce jitter to prevent stability issues related to nearly singular matrices, cf. Sect. 1.6.

The quality of the surrogate model directly correlates with the size of $\mathcal{D}$. A larger N typically ensures more accurate predictions but at the cost of increased training time. This dynamic presents a trade-off between computational efficiency and model accuracy. Lower-dimensional input spaces and narrower parameter ranges (for example, by choosing a Black-Scholes model instead of Heston) contribute to a denser, more effective training dataset, but at the potential cost of using a less appropriate model. A related consideration when using Monte Carlo simulation is an overall training budget. Recall Eq. (4.4) which would be repeated for all $(\mathbf{x}_i, \cdot) \in \mathcal{D}$, resulting in $N_{\text{tot}} = N \cdot \check{N}$ i.i.d. samples. Since the GP fitting time scales with N, one can improve the surrogate quality by instead increasing $\check{N}$ or budgeting additional simulations to key areas or of high certainty. Of course, N must be large enough so that $\mathcal{D}$ is representative of $\bar{\mathcal{D}}$, otherwise the interpolation may fail. It is also possible to have an adaptive approach where $\mathbf{x}$ is sequentially added to $\mathcal{D}$ based on some criterion until N_{tot} is depleted. This is done in [143] where tail risk measures (VaR and TVaR) are of interest, and explored in Sect. 5.2.

Extrapolation Since the GP model has no concept of the underlying structure, it will generally be a poor extrapolator beyond the training set due to reversion to the prior. One workaround is to use more elaborate prior mean functions as in Sect. 1.4.3 and/or composite kernels as in Sect. 2.3. For example, for many options the payoff is linear for extreme values, leading to a linear option value (e.g. value of Call or Put as a function of moneyness for $M \ll 1$ or $M \gg 1$). To preserve this linear property, one may utilize a a Matérn kernel and include a prior mean linear in M or use a compound kernel, adding a linear kernel to the Matérn. This compound kernel approach was utilized in [37] where the GP gives reasonable predictions outside the range of the training set, avoiding reversion to the prior that occurs without the Linear component. Another solution is to complement with synthetic "boundary" inputs to enforce out-of-sample extrapolation shape [1, 118]. To combat edge effects, i.e., worse performance around the boundaries of the training set, it is recommended to make the ranges of the training set slightly larger than the ranges of the actual estimation region.

Testing To evaluate the model's performance, one should similarly generate $\mathcal{D}_{\text{test}}$ which contains a collection of representative testing location. Generally speaking, the test set construction should mirror that of the training set, e.g. by sampling uniformly new inputs in each coordinate. If Monte Carlo is used, it is crucial to choose N ensure the $y_i \in \mathcal{D}_{\text{test}}$ are effectively noiseless, as they are considered the "truth" for model evaluation. Denoting $m(\mathbf{x})$ as the GP posterior mean as fitted to $\mathcal{D}$, the ith residual for $(\mathbf{x}_i, y_i) \in \mathcal{D}_{\text{test}}$ is $y_i - m(\mathbf{x}_i)$. From these residuals, common error metrics include root mean squared error (RMSE) L_2-norm, mean absolute error (MAE) L_1-norm, and maximum absolute error L_∞.

Fitting Market Options Data Instead of the modeller constructing $\mathcal{D}$, one may directly estimate the pricing functionals for option values based on historical option chains. This reflects the idea of incomplete information: for example the theoretical Call/Put strike-maturity surface is only observed at a few dozen liquidly traded strike-maturity combinations, and hence must be interpolated or inferred for further purposes. See e.g. Crépey and Dixon [37] who estimate equity option prices from historical observations of Call and Put prices, observed over four snapshots $t_1, \ldots, t_4$ in time with several hundred observations per snapshot. The training features are moneyness M, volatility σ, and time-to-maturity $\tau = T - t$. That study demonstrates the importance of including volatility as an input variable, in the sense of significantly higher GP posterior uncertainty if the volatility is excluded.

In this setting, two approaches for setting observation noise $\sigma(\mathbf{x})$ are possible. The first one considers the GP observations as being the mid-point prices $P_{\text{obs}} -$ $(P_{ask} + P_{bid})/2$, taking the observation noise variance to be proportional to the magnitude of the bid-ask spread. Maran and Pallavicini [125] use noise variance equal to one-fourth of the bid-ask spread squared, so $\sigma(\mathbf{x}) = (P_{ask}(\mathbf{x}) - P_{bid}(\mathbf{x}))/2$. The second one consists in considering both the bid and ask prices as two independent realizations of the Gaussian process, so that there are two measurements (i.e., replicated inputs) P_{bid}, P_{ask} for each input $\mathbf{x}$. The latter method will generally result in the GP option price surface f_* lying between the bid and ask surfaces. This gives two outputs per $\mathbf{x}_i$, meaning $\hat{\sigma}(\mathbf{x}_i)$ can be inferred using the methods in Sect. 3.1.

4.3 Option Sensitivities

The task of computing contract sensitivities is a fundamental problem in quantitative finance. Known as the *Greeks*, option sensitivities are essential for risk management and constructing dynamic hedges. *Delta hedging* manages risk by controlling for the sensitivity of the financial derivative to the underlying spot price; *Theta hedging* manages risk by controlling for the sensitivity to the passing of time, and so on.

Let $P(\mathbf{x})$ be the option price, where $\mathbf{x}$ is the current stochastic market state endowed with contract and model features as mentioned in Sect. 4.2. The Greeks estimation problem consists of evaluating the gradient ∂P with respect to the desired coordinate(s) of $\mathbf{x}$. Since those gradients are rarely available analytically, a strand of literature [27, 61, 146] addresses Greeks functional approximation.

Introduced in [37, 41] and expanded upon in [118], the idea is to fit a GP surrogate $\hat{P}$ to option prices, and then extract Greek estimates and their model-based uncertainty as the *gradient* of the GP. To fix ideas, consider a European option $P(\tau, S)$ depending on time-to-maturity $\tau = T - t$ and underlying price S, with the goal of learning its Delta

$$\Delta(\tau, S) \stackrel{\triangle}{=} \partial P(\tau, S)/\partial S.$$

We cast this objective in the framework of Sect. 4.1, starting with a training set $\mathcal{D} = \{(\mathbf{x}_i, y_i)\}_{i=1}^N$ of *prices* $y_i \simeq P(\mathbf{x}_i)$ and market factors $\mathbf{x}_i = (\tau_i, S_i)$, except the task is to learn $\mathbf{x} \mapsto \Delta(\mathbf{x})$. One advantage of the above strategy is that the GP surrogate $f(\cdot)$ has analytical derivatives (see Proposition 2.1 in Sect. 2.1), meaning all sensitivities are obtained in the same consistent manner as long as the parameters of interest are encoded into $\mathbf{x}$. For instance, to consistently extract Delta, Theta and Vega, one should train the GP surrogate on triples of time-to-maturity, asset price and asset volatility. The full GP surrogates are singularly suitable for this purpose thanks to their congruence with differentiation. To this end, we first provide a brief review of GP gradients.

4.3.1 GP Gradients

Starting with a prior GP $f \sim \mathcal{GP}(\mu, k)$, we consider the gradient of the resulting fitted GP model $f_* \sim \mathcal{GP}(m, s)$ with respect to the coordinate x_j. Applying Proposition 2.1 now to the posterior f_*, we have $\partial_{x_j} f_* \sim \mathcal{GP}(\partial_{x_j} m, \partial^2_{x_j, x_j'} s)$ with explicit formulae for its posterior mean and covariance in terms of the priors μ and k:

$$\partial_{x_j} m(\mathbf{x}_*) = \partial_{x_j} \mu(\mathbf{x}_*) + \partial_{x_j} k(\mathbf{x}_*, \mathbf{X}) \mathbf{K}_y^{-1} (\mathbf{y} - \boldsymbol{\mu}), \tag{4.6}$$

$$\partial^2_{x_j, x_j'} s(\mathbf{x}_*, \mathbf{x}_*') = \partial^2_{x_j, x_j'} k(\mathbf{x}_*, \mathbf{x}_*') - \partial_{x_j} k(\mathbf{x}_*, \mathbf{X}) \mathbf{K}_y^{-1} \partial_{x_j'} k(\mathbf{X}, \mathbf{x}_*'). \tag{4.7}$$

Thus, the gradient estimator (of coordinate x_j) is $\partial_{x_j} m(\mathbf{x}_*)$ in (4.6) which can be interpreted as formally differentiating the expression for $m(\cdot)$ with respect to $\mathbf{x}_j$:

$$\partial_{x_j} \mathbb{E}[f_*(\mathbf{x}_*)|\mathbf{X}, \mathbf{y}] = \partial_{x_j} \mu(\mathbf{x}_*) + \partial_{x_j} k(\mathbf{x}_*, \mathbf{X}) \mathbf{A}, \tag{4.8}$$

where $\mathbf{A} = \mathbf{K}_y^{-1}(\mathbf{y} - \boldsymbol{\mu})$ does not depend on the input [37]. The same formal method yields the posterior variance $\partial^2_{x_j, x_j'} s(\mathbf{x}_*, \mathbf{x}_*)$ of $\partial_{x_j} f_*$ in (4.7). The underlying structure is that differentiation is a linear operator that algebraically "commutes" with the Gaussian distributions defining a GP model. Consequently, one may iterate (by applying the chain rule further on $k(\cdot, \cdot)$ and its derivatives, provided they exist) to obtain analytic expressions for the mean and covariance of higher-order partial derivatives of f_*, yielding second-order and higher sensitivities.

The take-away is that we have analytic expressions for the posterior gradient GP $\partial_{x_j} f_*$. For example, Eqs. (4.6) and (4.7) can be used to construct credible bands for $\partial_{x_j} f_*(\mathbf{x})$ or even credible sets for e.g. $(f_*(\mathbf{x}), \partial_{x_j} f_*(\mathbf{x}))$.

Practically speaking, given a kernel k, to apply (4.6) we need to compute the various calculus derivatives of k. Below we show the resulting expressions for three common kernels discussed in Chap. 2. For the SE kernel (1.15) we have

$$\partial_{x_j} k_{SE}(\mathbf{x}, \mathbf{x}') = -\frac{x_j - x_j'}{\ell_{\text{len}, j}} k_{SE}(\mathbf{x}, \mathbf{x}'), \tag{4.9}$$

$$\partial^2_{x_j, x_j'} s^2(\mathbf{x}_*) = \frac{\eta^2}{\ell^2_{\text{len}, j}} - \partial_{x_j} k_{SE}(\mathbf{x}_*, \mathbf{X})(\mathbf{K}_{SE} + \sigma_\epsilon^2 \mathbf{I})^{-1} \partial_{x_j'} k_{SE}(\mathbf{X}, \mathbf{x}_*). \tag{4.10}$$

For the Matérn-5/2 kernel (2.11), we find

$$\partial_{x_j} k_{M52}(\mathbf{x}, \mathbf{x}') = \left(\frac{-\frac{5}{3\ell^2_{\text{len}, j}}(x_j - x_j') - \frac{5^{3/2}}{3\ell^3_{\text{len}, j}}(x_j - x_j')|x_j - x_j'|}{1 + \frac{\sqrt{5}}{\ell_{\text{len}, j}}|x_j - x_j'| + \frac{5}{3\ell^2_{\text{len}, j}}(x_j - x_j')^2} \right) k_{M52}(\mathbf{x}, \mathbf{x}'), \tag{4.11}$$

$$\partial^2_{x_j, x_j'} s^2(\mathbf{x}_*) = \frac{5\eta^2}{3\ell^2_{\text{len}, j}} - \partial_{x_j} k_{M52}(\mathbf{x}_*, \mathbf{X})(\mathbf{K}_{M52} + \sigma_\epsilon^2 \mathbf{I})^{-1} \partial_{x_j'} k_{M52}(\mathbf{X}, \mathbf{x}_*), \tag{4.12}$$

and for the Matérn-3/2 kernel (2.10):

$$\partial_{x_j} k_{M32}(\mathbf{x}, \mathbf{x}') = \left(\frac{-\frac{3}{\ell_{\text{len},j}^2}(x_j - x_j')}{1 + \frac{\sqrt{3}}{\ell_{\text{len},j}}|x_j - x_j'|} \right) k_{M32}(\mathbf{x}, \mathbf{x}'), \tag{4.13}$$

$$\partial_{x_j, x_j'}^2 s^2(\mathbf{x}_*) = \frac{3\eta^2}{\ell_{\text{len},j}^2} - \partial_{x_j} k_{M32}(\mathbf{x}_*, \mathbf{X})(\mathbf{K}_{M32} + \sigma_\epsilon^2 \mathbf{I})^{-1} \partial_{x_j'} k_{M32}(\mathbf{X}, \mathbf{x}_*). \tag{4.14}$$

Thus, after fitting a GP surrogate f_* to option prices, Greek estimation and model-based uncertainty quantification reduces to evaluating the formulas (4.6)–(4.7), e.g. based on (4.9), (4.11) or (4.13).

GP Gradients Through Finite-Difference Estimation
An alternative to obtain gradients is through a finite-difference estimator. For example, the second derivative of option price with respect to underlying is Gamma, $\Gamma(\tau, S) = \partial^2 P(\tau, S)/\partial S^2$. One may approximate Gamma using

$$P^{fd}(\tau, S; \delta) \triangleq \frac{P(\tau, S + \delta) - 2P(\tau, S) + P(\tau, S - \delta)}{\delta^2} \tag{4.15}$$

for a spatial discretization parameter $\delta > 0$. By using a GP surrogate f_* on the triplet of sites $\{(\tau, S - \delta), (\tau, S), (\tau, S + \delta)\}$, the resulting linear combination $f_*^{fd}(\tau, S; \delta)$ is Gaussian, yielding a predictive distribution for $\Gamma(\tau, S)$. The above formulas regarding gradients of $f_*(\cdot)$ (in this case, $\partial^2 f_*/\partial S^2$) are nothing else but the analytic limits of (4.15) as $\delta \to 0$.

4.3.2   Illustration: Estimating Greeks in the Black-Scholes Model

As illustration, we consider a Black–Scholes model with the goal of estimating the Delta, Theta and Gamma of Calls. This test environment provides an exact ground truth and hence ability to compute related errors by comparing the GP predictions to the classical Black–Scholes price formulae that yield closed-form expressions for the considered Greeks. Moreover, we can generate training data and observation noise of arbitrary amount and shape. We take bivariate input space $\mathbf{x} = (\tau, S)$ and use Monte Carlo to obtain training data $\mathbf{y}$. Namely, for each $y(\mathbf{x})$ we start with independent Monte Carlo simulations of size $\check{N} = 2500$ and then employ a plain sample average estimator as in (4.4) with the plug-in estimator $\hat{\sigma}(\mathbf{x}_i)$ for the input-dependent noise variance parameter as in (4.5). We use $r = 0.05$, $q = 0.01$, $\mathcal{K} = 100$, training on a two-dimensional set with $\tau \in [0, 0.6]$ and $S \in [60, 130]$. This case study is based on Ludkovski and Saporito [118]; see Dixon and Crepey [37] for a related application in the Heston model (4.3) where the GP is trained on the initial volatility v_0 and the goal is to estimate the Vega of the Call option (otherwise evaluated numerically via a COSINE solver).

In the illustrations below we choose to emphasize the impact of the GP kernel, which is the most important choice to be made by the user when computing GP sensitivities. Recall that different families imply different degree of smoothness in f_* and hence in the fitted Greeks. Figure 4.2 shows the fits and 95% credible bands across the SE and M52 kernel families and three different Greeks. While the training is done jointly in the τ and S dimension, we illustrate with one-dim plots that fix $\tau = 0.5$ and show dependence in S only. The top left panel shows the error $m(\tau, s) - P(\tau, S)$ between the predicted and true option prices, showing that both GP models perform well out-of-the-money and the largest error is in-the-money. This phenomenon is driven by the higher conditional variance, cf. (4.5) of training inputs Y_i in-the-money where Monte Carlo estimates (4.4) are less accurate. In essence, the observation noise is proportional to the price and hence estimating the latter is harder when $P(\tau, S)$ is higher. Otherwise, the top left panel echoes Fig. 4.1, showing that GP surrogates are excellent option price interpolators.

The top right panel displays the resulting estimate of the Delta $\widehat{\Delta}(\tau, \cdot)$'s which are the gradients of the respective surrogates $\partial_S f_*(\tau, S)$. Both kernels do well in matching the true Delta, though the M52 fit exhibits some fluctuations on the right of the plot hinting at some instability of the respective estimated gradient. The bottom left panel illustrates the fitted $\widehat{\Theta}(\tau, \cdot)$ which uses the exact same GP models as in the first row of the figure, simply evaluating the gradient $\partial_\tau f_*(\tau, S)$. Due to the more complex shape of the Theta, and in particular higher convexity of $P(\tau, S)$ in τ, the quality of $\widehat{\Theta}$ is somewhat worse compared to that of Delta. In particular, the SE surrogate underestimates the steep trough of Θ ATM, estimating $\partial_\tau m(0.1, 100) \simeq -7.2$ rather than the true -7.6. In contrast, the M52 fit is more flexible and performs better. Finally, the bottom right panel of Fig. 4.2 illustrates the learning of $\Gamma(\tau, S)$. Here, we use the finite-differencing approximation in (4.15). Numerical estimation of second-order sensitivities is extremely challenging, especially through functional approximators. We observe that the Matérn-5/2 kernel exhibits obvious numeric instability in evaluating (4.15), in part because it is only twice differentiable. As a result, the fit using k_{M52} is likely to exhibit high-order fluctuations; these oscillations can also be seen in the k_{M52}-based Delta deep-in-the-money ($S \gg 110$). The SE surrogate performs really well given that it was trained on just 400 noisy observations; this behavior is linked to its tendency to over-smooth spatially.

A related issue for Greeks estimation is the reversion to the prior. Without any adjustments, the GP estimates of the gradient are poor around the edges of the training set. For instance, reversion to a constant prior mean would cause $\widehat{\Delta}(\tau, S) < 1$ at the right edge and $\widehat{\Delta}(\tau, S) > 0$ at the left edge. In the shown example, this is mitigated by imposing a "boundary condition": virtual training

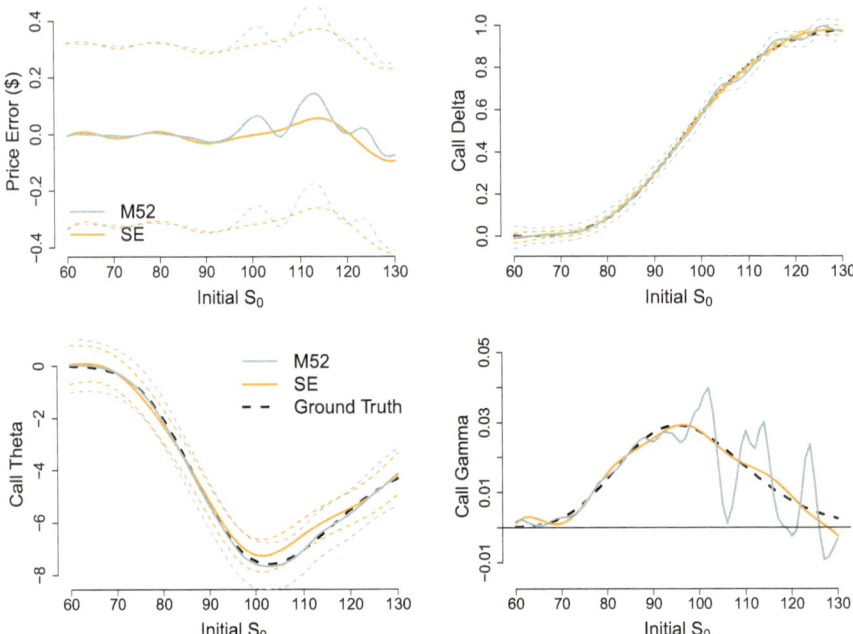

Fig. 4.2 Estimated sensitivities (GP posterior mean) at $\tau = 0.5$ together with 95% credible intervals for a Black–Scholes Call learned via a GP with SE and M52 kernels and using a space-filling training design. The Gamma is computed using the finite difference approximation (4.15) on f_*. Ground truth indicated in dashed black line. Training set of $N = 400$ inputs and $\check{N} = 2500$ inner Monte Carlo simulations

points that enforce $\widehat{\Delta}(\tau, S) \simeq 1$ for $S \simeq 130$ and $\widehat{\Delta}(\tau, S) \simeq 0$ for $S \simeq 60$, see Sect. 4.2 and [118]. Other solutions are to impose informative prior means, or train on a sufficiently large input domain to resolve edge and extrapolation effects.

The different kernels also yield different credible bands. The bands of the k_{SE}-surrogate are narrower compared to the M52. This matches the stylized fact that the smoother $m(\cdot)$, the tighter the CI; thus for instance the bands of a M32 surrogate would be even wider [118]. In the example, the bands of the SE surrogate are generally *too* narrow, i.e., its posterior uncertainty is underestimated. Consequently, coverage levels (the frequency of the ground truth belonging to the estimated 95% CI) are below the nominal 95% level. In contrast, the M52 surrogate is better calibrated to judge its own accuracy. The above discussion highlights that there is no clear-cut winner between the two shown GP fits and a more nuanced and context-driven analysis is warranted.

The squared-exponential (SE) kernel will lead to very smooth fits, while Matérn kernels allow more roughness, controlled by v. In turn, smoother kernels yield narrower uncertainty on the gradient, while rougher kernels will create wider confidence intervals.

Remark 4.1 By default, the GP surrogate is not aware of any no-arbitrage or other structural constraints. Therefore, the Greeks predictions can be financially nonsensical, such as predicting $\widehat{\Delta}(\tau, S) > 1$ which occurs (just barely) in Fig. 4.2 around $S = 120$. This can be addressed using constrained GPs discussed in the next Section, or via gradient observations, cf. (4.18).

4.4 Constrained GPs and No-Arbitrage

Option pricers must satisfy certain model-free constraints imposed by *no-arbitrage*. This refers both to smoothness constraints, e.g. that the price surface should be differentiable (important for deriving the local volatility surface), as well as shape constraints. For example, if $P(\cdot)$ is a Call option no-arbitrage implies:

- $P(\cdot)$ is monotonically increasing in time-to-maturity τ (Calendar spread constraint);
- $P(\cdot)$ is monotonically increasing in moneyness $M = S/\mathcal{K}$ (Bull spread constraint);
- $P(\cdot)$ is convex in moneyness M (Butterfly spread constraint).

These constraints can be coded as being "soft", e.g. through adding penalty functions to the loss function to encourage the desired sign of the derivatives of f, or as "hard" constraints, enforcing the restrictions explicitly.

Adding positivity constraint and several others (such as monotonicity, convexity, boundedness) for GPs is nontrivial since the posterior process is no longer Gaussian. Moreover, the desired constraints are usually infinite-dimensional as they must hold everywhere in the state space. Multiple different techniques have been proposed. The frameworks presented in [73, 142, 166] impose the constraints only at a subset of "virtual" input locations, without guaranteeing that the constraints hold in the entire domain. For instance, in incorporating monotonicity constraints, [166] proposed to select the locations with large probabilities of having negative derivatives as the virtual locations set in which they enforce the constraints. The approach of [4] extended this for multiple constraints through an efficient method for sampling the posterior process based on the derivation of the posterior of the constrained Gaussian process using a linear operator. Maatouk and Bay [121] propose a finite dimensional approximation of Gaussian processes for which the inequality constraints are easy to check and can be guaranteed in the whole domain. Note that these methods generally rely on efficient sampling from a truncated multivariate Gaussian distribution, needed to estimating the posterior process and for sampling from the posterior constrained GP; a common approach is the Hamiltonian Monte Carlo scheme in [108].

Below we summarize the strategy in Cousin et al. [34] which consists in building a finite-dimensional GP to impose no-arbitrage constraints on the strike-maturity bivariate option surface of a European call option $P(T, \mathcal{K})$. Consider a grid $\mathcal{G} =$

$\{(T_i, \mathcal{K}_j) : i = 1, \ldots, N_T, j = 1, \ldots, N_K\}$ with equal spacing $h_T, h_{\mathcal{K}}$ and $N_{\mathcal{G}} = N_T \cdot N_K$ total locations. Construct the bivariate *hat* function

$$\phi_{i,j}(T, \mathcal{K}) \stackrel{\triangle}{=} \max\left(1 - \frac{|T - T_i|}{h_T}, 0\right) \cdot \max\left(1 - \frac{|\mathcal{K} - \mathcal{K}_j|}{h_{\mathcal{K}}}, 0\right). \tag{4.16}$$

Endowing the prior $f \sim GP(\mu, k)$, linearity ensures the finite dimensional approximation is another GP $\bar{f}$:

$$\bar{f}(T, \mathcal{K}) \stackrel{\triangle}{=} \sum_{i=1}^{N_T} \sum_{j=1}^{N_K} \phi_{i,j}(T, \mathcal{K}) f(T_i, \mathcal{K}_j). \tag{4.17}$$

The approximation $\bar{f}$ is just a linear interpolation of f based on its values at the grid points: at most four of the $\phi_{i,j}$ terms are non-zero for a given $(T, \mathcal{K})$, namely its nearest neighbors on the grid. Equation (4.17) can be explicitly written as the dot product $\bar{f}(T, \mathcal{K}) = \boldsymbol{\phi}(T, \mathcal{K})^\top \mathbf{f}$ where $\boldsymbol{\phi}(T, \mathcal{K}) \in \mathbb{R}^{N_{\mathcal{G}}}$ rasterizes the $N_T \times N_K$ matrix with i, j entry $\phi_{i,j}(T, \mathcal{K})$, and $\mathbf{f} \in \mathbb{R}^{N_{\mathcal{G}}}$ is the Gaussian vector from f over $\mathcal{G}$ with indices matching $\boldsymbol{\phi}$, so its $N_{\mathcal{G}} \times N_{\mathcal{G}}$ covariance matrix has entries $\mathbf{K}_{ij,i'j'} = k((T_i, \mathcal{K}_j), (T_{i'}, \mathcal{K}_{j'}))$. Moreover, monotonicity/convexity constraints on $\bar{f}$ translate into those on the finite $\mathbf{f}$. For example,

- $\bar{f}(T, \mathcal{K})$ is monotone increasing in T if and only if $f(T_i, \mathcal{K}_j) \geq f(T_{i+1}, \mathcal{K}_j)$ for all i, j.
- $\bar{f}(T, \mathcal{K})$ is convex in $\mathcal{K}$ if and only if $f(T_i, \mathcal{K}_{j+2}) - f(T_i, \mathcal{K}_{j+1}) \geq f(T_i, \mathcal{K}_{j+1}) - f(T_i, \mathcal{K}_j)$ for all i, j.

Therefore, the global constraints on $\bar{f}(\cdot)$ become a system of inequalities C_{ineq} on the finite Gaussian vector $\mathbf{f}$. Thus, the regression assumption $y = \bar{f}(T, \mathcal{K}) + \epsilon$ with constraints on $\bar{f}$ is equivalent to $y = \boldsymbol{\phi}(T, \mathcal{K})^\top \mathbf{f} + \epsilon$ with $\mathbf{f} \in C_{\text{ineq}}$. With N training observations, this means $\mathbf{y} = \boldsymbol{\Phi}^\top \mathbf{f} + \boldsymbol{\epsilon}$, where $\boldsymbol{\Phi}^\top$ is $N \times N_{\mathcal{G}}$ with rows given by $\boldsymbol{\phi}(T, \mathcal{K})^\top$. Given hyperparameters defining $\mathbf{f}$, the constrained MAP $\mathbf{m}_c$ can be obtained by maximizing the conditional density of $\mathbf{f}|\mathbf{f} \in C_{\text{ineq}}$, equivalent to the solution coming from the quadratic problem [34]

$$(\mathbf{m}_c, \boldsymbol{e}_c) = \arg\min_{\mathbf{y} = \boldsymbol{\Phi}^\top \mathbf{m} + \boldsymbol{e}, \mathbf{m} \in C_{\text{ineq}}} \left(\mathbf{m}^\top \mathbf{K}^{-1} \mathbf{m} + \boldsymbol{e}^\top \Sigma_\epsilon^{-1} \boldsymbol{e}\right)$$

Consequently, the MAP $\bar{m}_c(T, \mathcal{K})$ of the constrained $\bar{f}(T, \mathcal{K})$ is given by (4.17) with $m_c(T_i, \mathcal{K}_j)$ replacing $f(T_i, \mathcal{K}_j)$. Posterior samples of $\bar{f}$ are similarly obtained by utilizing the standard GP equations but truncating to C_{ineq}, initialized with $\bar{m}_c$ to ensure constraints are satisfied.

Note that [34, 125] first obtain hyperparameters to maximize the likelihood of $\mathbf{y} \sim MVN(\mathbf{0}, \boldsymbol{\Phi}^\top \bar{\mathbf{K}} \boldsymbol{\Phi} + \Sigma_\epsilon)$. A full example is presented in Cousin and Gueye [34] working with a grid of size 600 with $N_T = 30$, $N_K = 20$ based on fitting Put and Call prices on Euro Stoxx 50 futures using a separable SE kernel over T and $\mathcal{K}$ dimensions.

Derivative Observations

It is possible to generalize the above results to consider the full joint distribution of $[f, \nabla f]$ based on the fact that differentiation is a linear operator, so the latter is a (bi-output) GP. In turn, this allows to consider observations of ∇f as part of the training data, i.e., to include information about gradients of the response. To ease notation, let $g \sim \mathcal{GP}(\mu_g, k_g)$ be the gradient of f (viewed as a d-vector at a given $\mathbf{x}$), observed as (a vector) y_g with noise $\sigma_g(\mathbf{x}) \geq 0$, $y_g(\mathbf{x}_i) = \nabla f(\mathbf{x}_i) + \epsilon_g(\mathbf{x}_i)$. Using exactly the same logic as in (1.7) leads to the joint covariance of $[\mathbf{y}, \mathbf{y}_g]$ as the $(N+dN) \times (N+dN)$ block matrix

$$\tilde{\mathbf{K}} \triangleq \begin{bmatrix} K_f(\mathbf{X}, \mathbf{X}) & (\nabla \mathbf{K})^\top \\ \nabla \mathbf{K} & \nabla^2 K_f(\mathbf{X}, \mathbf{X}) \end{bmatrix} + \begin{bmatrix} \Sigma_f(\mathbf{X}, \mathbf{X}) & 0 \\ 0 & \Sigma_g(\mathbf{X}, \mathbf{X}) \end{bmatrix}$$

where $\nabla \mathbf{K} \triangleq \nabla K(\mathbf{X}, \mathbf{X})$ is the $(dN) \times N$ matrix of the $\partial_{x_d} k(\mathbf{x}_i, \mathbf{x}_j)$ and Σ_g is the $(dN) \times (dN)$ diagonal matrix of $\sigma_g(\mathbf{x}_i)$'s.

Therefore, reusing the methods of Chap. 1 (or more precisely the multi-output equations), given observations $\mathbf{y}_g$ we have the joint posterior mean

$$\begin{bmatrix} m_*(\mathbf{x}_*) \\ g_*(\mathbf{x}_*) \end{bmatrix} = \begin{bmatrix} \mu(\mathbf{x}_*) \\ \mu_g(\mathbf{x}_*) \end{bmatrix} + \begin{bmatrix} K_f(\mathbf{x}_*, \mathbf{X}) & (\nabla \mathbf{k}_*)^\top \\ \nabla \mathbf{k}_* & \nabla^2 k(\mathbf{x}_*, \mathbf{X}) \end{bmatrix} \tilde{\mathbf{K}}^{-1} \left(\begin{bmatrix} \mathbf{y} \\ \mathbf{y}_g \end{bmatrix} - \begin{bmatrix} \mu(\mathbf{X}) \\ \mu_g(\mathbf{X}) \end{bmatrix} \right)$$

$$(4.18)$$

where $\nabla \mathbf{k}_*$ is the $dN \times 1$ vector of the $\partial_{x_d} k(\mathbf{x}_i, \mathbf{x}_*), i = 1, \ldots, N$. If observations at some inputs are missing some of the gradients, or have no gradient observations, or have gradients observed but no $y(\mathbf{x}_i)$ itself, one may simply delete the corresponding rows/columns in $\tilde{\mathbf{K}}$, $\nabla \mathbf{k}_*$ and Σ matrices.

More generally, one can work with any linear operators $\mathcal{L}$. The mean and covariance of $\mathcal{L}f$ is $\mathcal{L}\mu$ and $\mathcal{L}^2 k$, for example $\mathcal{L}f(x) = \partial_j f(x)$ leads to $\mathcal{L}^2 k(x, x') = \frac{\partial^2 k}{\partial x_j^2}(x, x')$, collapsing to $\mathcal{L}^2 k = \nabla^2 k$. Beyond gradients, this formalism allows for instance to handle integrals of f, useful in Chap. 6, or to consider observations via directional derivatives in the multivariate case (i.e., a dot product of the gradient ∇f with a fixed vector u).

4.5   Portfolio Modeling and Credit Valuation Adjustments

Given a collection of financial instruments $P^{(\ell)}$, $\ell = 1, \ldots, L$ to be priced, one approach is to separately build L GP models for each one and then evaluate the portfolio as the linear combination of the independent surrogates: $\pi = \sum_{\ell=1}^{L} w_\ell \widehat{P}^{(\ell)}$. An alternative proposed by [37] is to train a multi-output GP to $\mathbf{P} = [P^{(1)}, \ldots, P^{(L)}]^\top$, cf. Sect. 3.3. One motivation is to account for the dependence between instruments that occurs due to shared (or correlated) risk factors. For example, different Calls and Puts on the same underlying clearly share a common dependence on asset price S. A MOGP is able to quantify the resulting posterior cross-covariance of the GP predictions and moreover borrow additional information from the ℓ-th instrument observation to help with learning the L-output functional. Hence, with correlated risks there is a conceptual preference for MOGP as the surrogate for portfolio analysis. Note that often the cross-covariance is negative (e.g. a Call is positively correlated to S but a Put is negatively correlated), so the ICM structure (3.19) might not be appropriate.

A common application is in *Credit Valuation Adjustments (CVA)*, where one computes the expected cost of counterparty risk, or equivalently, the expected loss from recovery on the market value of the portfolio in case of counterparty default. Given a bilateral portfolio with net value $\pi_t(\mathbf{X}_t)$ where $\mathbf{X}_t$ denotes market factors (emphasizing dependence on t), counterparty default time τ with CDF $F_\tau(\cdot)$, and horizon T, the CVA at time zero is given by [37]

$$CVA_0 = (1 - R) \int_0^T e^{-rt} \mathbb{E}[\max(\pi_t(\mathbf{X}_t), 0)] dF_\tau(t),$$

where R is the recovery rate in the event of default, and $\mathbb{E}$ is the credit valuation measure. Assuming there is a stochastic intensity γ_t for τ, we may re-write as $CVA_0 = (1 - R)\mathbb{E}[\int_0^T e^{-rt} \max(\pi_t(\mathbf{X}_t), 0)e^{-\int_0^t \gamma_s ds} \gamma_t dt]$. The standard approach to evaluate CVA_0 is to approximate with a double Monte Carlo simulation, averaging M paths with market scenarios $\mathbf{x}_{\cdot, m}$ across time steps $t_k = k\Delta t$ to account for default time:

$$CVA_0 \simeq \frac{(1 - R)h_t}{M} \sum_{m=1}^{M} \sum_{k=1}^{K} \max\left(\pi_{t_k}(\mathbf{x}_{t_k, m}), 0\right) e^{-rt_k - h_t \sum_{i<k} \gamma(t_i, \mathbf{x}_{t_i, m})} \gamma(t_k, \mathbf{x}_{t_k, m})$$

$$(4.19)$$

The nested simulation (4.19) requires $M \cdot K$ evaluations of the portfolio value $\pi_{t_k}(\mathbf{x}_{t_k, m})$, which often involve exotic e.g. path-dependent options. Recalling Eq. (4.4), a naive implementation is computationally prohibitive, requiring $\check{N} \cdot M \cdot K$ simulations. Dixon and Crepey [37] replace $\pi_{t_k}(\mathbf{x}_{t_k, m})$ with a GP surrogate $\hat{\pi}(\mathbf{x})$ (noting that t is part of $\mathbf{x}$ as π is Markovian) trained according to a size-N training set $\mathcal{D}$ as described in Sect. 4.2, saving one level of nested revaluation. In addition to improving computation time, it provides a probabilistic surrogate of $\pi_{t_k}(\mathbf{x}_{t_k,m})$. See the full work of [37] which adapts this to one-year CVA Value-at-Risk with uncertainty quantification.

Further Reading

Option pricing surrogates have been one of the key impetuses for connecting machine learning and quantitative finance. Multiple approaches, including with commercial success [1, 57], have emerged based on neural network surrogates or gradient boosting [40]. We refer the readers to the monograph by Dixon et al. [48] and the handbook [26] on broader summaries and comprehensive surveys of machine learning approaches to option pricing and Greek estimation. For applications of GPs, the key references we have relied on are [37, 41, 118]. See also the survey [114].

Shape-constrained GPs constitute an active area of methodological research and are important for financial contexts. See the survey [158], or the lineqGPR Python package based on the works [9, 109].

This Chapter includes an R-based notebook that illustrates the learning of option prices and Deltas within the Black-Scholes and Heston models, reproducing Figs. 4.1 and 4.2.

Chapter 5
Optimal Stopping

This Chapter discusses applications of GPs to optimal stopping problems, especially pricing of American-style options. We consider the use of GP surrogates to learn continuation values connecting to the Regression Monte Carlo framework for simulation-based solvers of optimal stopping.

In an optimal stopping problem (OSP), the goal is to find a random time τ to maximize the corresponding payoff $G(\mathbf{X}_\tau)$ where $(\mathbf{X}_t)_{t \geq 0}$ is the stochastic state process with state space $\mathcal{X} \subseteq \mathbb{R}^d$.[1] The intuition is that the payoff evolves randomly according to the dynamics of $(\mathbf{X}_t)$ and the goal is to time the decision τ to collect the payoff at the most profitable (on average) time. A fundamental idea is that decisions must be done on-the-fly, without any hindsight. Mathematically, τ must be a stopping time with respect to a given filtration $\mathbb{F} = (\mathcal{F}_t)$, typically the natural filtration generated by $(\mathbf{X}_t)$ as mentioned at the beginning of Chap. 4. As such, the decision to stop at t can only be based on information in $\mathcal{F}_t$, without any advance knowledge of the future. In order to apply surrogate methods, we also assume that $(\mathbf{X}_t)$ is rich enough so that the Markov property applies. Further, $\mathbf{X}_t$ may be augmented as in Chap. 4 with non-stochastic quantities (e.g. model and market related) to admit related analysis.

The motivating example of OSP in finance is valuing American-style contracts, that allow the buyer to exercise any time on the interval $[0, T]$ with (discounted) payoff $G(t, \mathbf{X}_t)$ if exercised at time t. The goal is then to evaluate the *value function* $V : [0, T] \times \mathcal{X} \to \mathbb{R}$ representing maximum future expected reward:

$$V(t, \mathbf{x}) \stackrel{\triangle}{=} \sup_{\tau_t \in \mathfrak{S}_t} \mathbb{E}^Q \left[G(\tau_t, \mathbf{X}_{\tau_t}) \mid \mathbf{X}_t = \mathbf{x} \right], \tag{5.1}$$

---

[1] In this chapter we use subscripts for time indexing and superscripts to index training inputs.

M. Ludkovski, J. Risk, *Gaussian Process Models for Quantitative Finance*,
SpringerBriefs in Quantitative Finance,
https://doi.org/10.1007/978-3-031-80874-6_5

where $\mathfrak{S}_t$ is the set of $\mathbb{F}$-stopping times valued in $[t, T]$. The prototypical example is a Bermudan Put with payoff $G(t, x) = e^{-rt}(K - x)_+$, where the contract buyer must decide, non-anticipatively, on the best stopping time $\tau \leq T$ to exercise her Put. In turn, standard hedging arguments (assuming a complete market with risk-neutral measure Q) imply that $V(t, \mathbf{x})$ is the no-arbitrage contract value at time t given initial market state $\mathbf{x}$.

For the remainder of this Chapter we adopt the discrete-time paradigm of Bermudan options that allow for exercise at a predetermined collection of K (discrete) times $\mathcal{T} = \{t_k : k = 0, 1, 2, \ldots, K, \ t_K = T\}$ up to T. These exercise times are equally spaced and denoted by $t_k = k \cdot T/K$, although the proceeding discussion extends naturally to the general discrete case. For ease of exposition we substitute t_k with k when contextually unambiguous. Therefore, $\tau_k^* \in \mathfrak{S}_k$ is optimal if $V(k, \mathbf{x}) = \mathbb{E}[G(\tau_k^*, \mathbf{X}_{\tau_k^*})| \ \mathbf{X}_k = \mathbf{x}]$ for all $\mathbf{x}$; the sequence $(\tau_k^*)_{k=0}^T$ is an *optimal stopping rule*.

To solve the discrete-time OSP, we recall the following approach based on the dynamic programming principle (DPP). Define the *continuation value*

$$q(k, \mathbf{x}) \stackrel{\triangle}{=} \sup_{\tau_{k+1} \in \mathfrak{S}_{k+1}} \mathbb{E}\left[G(\tau_{k+1}, \mathbf{X}_{\tau_{k+1}})| \ \mathbf{X}_k = \mathbf{x}\right] = \mathbb{E}[V(k + 1, \mathbf{X}_{k+1})| \ \mathbf{X}_k = \mathbf{x}], \tag{5.2}$$

where the latter equality follows from the tower property of conditional expectations applied to (5.1). The continuation value represents the optimal value at time t if one is forced to continue. Note that the value function and continuation value are related by $V(k, \mathbf{x}) = \max(G(k, \mathbf{x}), q(k, \mathbf{x}))$. The DPP asserts a recursive solution to the OSP for $k = T - 1, \ldots, 0$ according to $V(K, \mathbf{X}_K) = \max(G(k, \mathbf{X}_K), 0)$, $q(K, \mathbf{X}_K) = 0$ and

$$
\begin{aligned}
V(k, \mathbf{X}_k) &= \max(G(k, \mathbf{X}_k), q(k, \mathbf{X}_k)), \\
q(k, \mathbf{X}_k) &= \mathbb{E}[\max\left(G(k + 1, \mathbf{X}_{k+1}), q(k + 1, \mathbf{X}_{k+1})\right)| \ \mathbf{X}_k].
\end{aligned}
\tag{5.3}
$$

Similarly, the optimal stopping times τ_k^* can be derived through

$$
\begin{aligned}
\tau_k^* &= k \mathbb{1}_{\{V(k, \mathbf{X}_k) = G(k, \mathbf{X}_k)\}} + \tau_{k+1}^* \mathbb{1}_{\{V(k, \mathbf{X}_k) > G(k, \mathbf{X}_k)\}} \\
&= k \mathbb{1}_{\{q(k, \mathbf{X}_k) \leq G(k, \mathbf{X}_k)\}} + \tau_{k+1}^* \mathbb{1}_{\{q(k, \mathbf{X}_k) > G(k, \mathbf{X}_k)\}},
\end{aligned}
\tag{5.4}
$$

noting that the decision to continue is governed by checking whether the expected reward-to-go dominates the immediate payoff.

Making connections with (5.4), the OSP can be seen as simplified version of dynamic decision making, providing an intuitive alternative to an approach based purely through (5.3). This can be made precise by re-writing the decision to continue as a function of k and current state $\mathbf{x} \in \mathcal{X}$, represented as an *action map* $A_k(\mathbf{x}) \in \{0, 1\}$ where 0 corresponds to stopping in state $\mathbf{x}$ at step k and 1 to continuing. The sequence $A_0(\cdot), A_1(\cdot), \ldots, A_T(\cdot)$ induces a corresponding sequence of first hitting times $(\tau_{A_{k:T}})_{k=0}^T$ given by

$$\tau_{A_{k:T}} \triangleq \min\{\ell \geq k : A_\ell(\mathbf{X}_\ell) = 0\} \wedge T \tag{5.5}$$

$$= \sum_{m=k}^T m \cdot (1 - A_m(\mathbf{X}_m)) \prod_{\ell=k}^{m-1} A_\ell(\mathbf{X}_\ell).$$

From (5.4), the optimal action map is defined by $A_k^*(\mathbf{x}) = 1_{\{q(k,\mathbf{X}_k)>G(k,\mathbf{X}_k)\}} = 1_{\{V(k,\mathbf{X}_k)>G(k,\mathbf{X}_k)\}}$, and so the induced $(\tau_{A_{k:T}^*})_{k=0}^T$ coincides with the optimal stopping rule $(\tau_k^*)_{k=0}^T$. One benefit of this perspective is that solving an OSP is equivalent to optimally classifying a given $(k, \mathbf{x})$ into an equivalent *stopping region* or *action region* $S_k \triangleq \{\mathbf{x} \in X : A_k(\mathbf{x}) = 0\} \subseteq X$ or its complement, the *continuation set*. Connecting to our running example of a 1d Bermudan Put, it is well known that for a given k, $S_k^* = [0, \bar{s}_k]$, meaning one should stop as soon as the asset price drops below a certain exercise threshold $\bar{s}_k$ depending on k. Thus, for $x > \bar{s}_k$ we have the optimal $A_k^*(x) = 1$ (continue), but as soon as $x \leq \bar{s}_k$ then $A_k(x) = 0$ (stop). Consequently, $A_k^*(x) = 1_{\{x>\bar{s}_k\}}$ defines the optimal $\tau_{A_{k:T}^*} = \min\{\ell \geq k : X_\ell \leq \bar{s}_\ell\}$ to be the first time the state crosses the exercise *boundary* $\bar{s}$..

5.1 Regression Monte Carlo

Moving toward numeric implementation, we work with the *timing value*,

$$T(k, \mathbf{x}) \triangleq q(k, \mathbf{x}) - G(k, \mathbf{x})$$

which abstracts from the payoff function specifics. Thus, exercising is optimal when $T(k, \mathbf{x}) < 0$, and the stopping boundary is the zero contour of $T(k, \cdot)$. Given $T(k, \mathbf{x})$ one can recover the value function (and hence the option price) via $V(k, \mathbf{x}) = G(k, \mathbf{x}) + \max(0, T(k, \mathbf{x}))$. Financially, a positive (negative) timing value implies a positive (zero) early exercise premium.

The Regression Monte Carlo (RMC, a specific case being the Longstaff Schwartz Algorithm or Least Squares Monte Carlo) framework [53, 107] approximates the OSP by constructing a strategy $(\widehat{A}_0, \widehat{A}_1, \ldots, \widehat{A}_T)$. This is done through the backward recursion mentioned above, generating functional approximations of the timing values $\widehat{T}(k, \cdot)$ for each k in order to build the action maps $\widehat{A}_k(\cdot)$. Specifically, the RMC loop is initialized with $\widehat{V}(T, \mathbf{x}) = G(T, \mathbf{x})$, and recursively for $k = T - 1, \ldots, 1, 0$:

$$\begin{cases} \text{learn the timing value } \widehat{T}(k, \cdot) = \widehat{q}(k, \cdot) - G(k, \cdot), \text{ and set} \\ \widehat{A}_k(\cdot) \triangleq 1_{\{\widehat{T}(k,\cdot)>0\}}, \\ \widehat{V}(k, \cdot) \triangleq G(k, \cdot) + \max(0, \widehat{T}(k, \cdot)). \end{cases} \tag{5.6}$$

Note that dependence on the previous $\widehat{V}(k+1, \cdot)$, $\widehat{A}_{k+1}, \ldots, \widehat{A}_{T-1}$ only exists in the principal step of learning $\widehat{T}(k, \cdot)$ (equivalently, $\widehat{q}(k, \cdot)$) and can be done by learning one of the conditional expectation mappings

$$\mathbf{x} \mapsto \mathbb{E}[\widehat{V}(k+1, \mathbf{X}_{k+1})|\mathbf{X}_k = \mathbf{x}]; \text{ or} \tag{5.7}$$

$$\mathbf{x} \mapsto \mathbb{E}[G(\tau_{\widehat{A}_{k+1:T}}, \mathbf{X}_{\tau_{\widehat{A}_{k+1:T}}})|\mathbf{X}_k = \mathbf{x}] \tag{5.8}$$

where in both cases the expectation is over $(\mathbf{X}_s)_{s>k}$. The former is the Tsitsiklis and Van Roy [161] algorithm and the latter the Longstaff and Schwartz [107] version. See also a multi-step look-ahead version in [54] which subsumes the two. While learning (5.7) is numerically simpler, (5.8) incorporates all time points up to the final T, which may be important depending on the distribution of $(\mathbf{X}_s)_{s>k}|\mathbf{X}_k = \mathbf{x}$ and the collection of training designs.

As a final step, one can compute the initial option value $\widetilde{V}(0, \mathbf{x})$ as the learned mapping $\mathbb{E}[\widehat{V}(1, \mathbf{X}_1)|\mathbf{X}_0 = \mathbf{x}]$ if using (5.7). When using (5.8), the collection of fitted action maps induce the expected reward

$$\widetilde{V}(0, \mathbf{x}; \widehat{A}_{0:T}) \triangleq \mathbb{E}\left[G(\tau_{\widehat{A}_{0:K}}, \mathbf{X}_{\tau_{\widehat{A}_{0:K}}})\Big| \mathbf{X}_0 = \mathbf{x}\right]. \tag{5.9}$$

The gap between $\widetilde{V}(0, \mathbf{x})$ and $V(0, \mathbf{x})$ (which is the maximal possible expected reward) is the performance metric for evaluating $\widehat{A}_{0:K}$.

The core step of fitting $\widehat{T}(k, \cdot)$ connects to the surrogate modeling theme in this book. Probabilistically, the timing value is a conditional expectation, i.e. the expected response from a stochastic simulator. Financially, timing value measures the relative gain from not exercising. The RMC paradigm translates the above into statistical language, reinterpreting the timing value for a given step k as the empirical minimizer in some function space $\mathcal{H}_k$ of the mean squared error from the observations,

$$\widehat{T}(k, \cdot) \triangleq \arg\min_{g \in \mathcal{H}_k} \sum_{n=1}^{N_k} (g(\mathbf{x}_{n,k}) - y_n)^2. \tag{5.10}$$

The classical RMC approach employs the L_2 loss function for each k with user-specified basis functions $\mathcal{H}_k$, fitted based on a training set $\mathcal{D}_k$. The "Monte Carlo" reference is to the standard strategy of constructing $\mathcal{D}_k$ from M i.i.d. *paths* of $(\mathbf{X}_k)_{k\geq 0}$ (thus $\mathcal{D}_k$ and $\mathcal{D}_{k+1}$ are not independent). In order to obtain more flexible "non-parametric" fits, many other approaches, for example splines or neural networks [97, 98] (as well as the earliest related publication [28] by Carriére who considered kernel regression) have been proposed to obtain $\widehat{T}$'s. Gaussian Process surrogates for RMC were originally discussed in [111] and also investigated in [75]. With GPs, the objective function in (5.10) changes to the RKHS regularized solution (see Eq. (1.31)), resulting in the posterior mean estimate $\widehat{T}(k, \cdot) = m_k(\cdot)$.

Advantageously, the GP gives a full distributional response beyond something like a traditional kernel ridge regressor. As we will see in the next section, the posterior standard deviation $s_k(\cdot)$ enables a method of sequential design. Another advantage of modeling $\widehat{T}$ via GP is the ability to fit complex input-output relations based on just a few training inputs.

The basic construction for a GP surrogate for RMC is essentially the same as the previous recipe as in Sect. 4.2, modified to handle the sequential nature. The basic template requires $K = T/\Delta t$ designs $\mathcal{D}_0, \dots, \mathcal{D}_{K-1}$ with N_k observations in each $\mathcal{D}_k$. For each input $\mathbf{x} \in \mathcal{D}_k$, a trajectory of the future $(\mathbf{X}_s)_{s>k}$ given $\mathbf{X}_k = \mathbf{x}$ is generated to return the realized sample $(\mathbf{x}_s)_{s>k}$ and subsequently $y(\mathbf{x})$, e.g. $\hat{V}(k+1, \mathbf{x}_{k+1})$ or $G(\tau_{\widehat{A}_{k+1:T}}, \mathbf{x}_{\tau_{\widehat{A}_{k+1:T}}})$, differenced by $G(k, \mathbf{x})$. Regardless, $y(\mathbf{x})$ is a random (i.e. pathwise) realization of the timing value $T(k, \mathbf{x})$. This fully populates $\mathcal{D}_k$, so that a resulting GP $f_k | \mathcal{D}_k$ is obtained for each k with $m_k(\cdot) = \widehat{T}(k, \cdot)$. Recursively, (5.6) yields $\widehat{T}(k, \cdot)$, $\widehat{A}_k(\cdot)$, and $\widehat{V}(k, \cdot)$ for all $k = K - 1, \dots, 1, 0$.

5.1.1 RMC GP Features

A running theme in this book is highlighting the specific features of an application to pinpoint what aspects of the GP model are critical for successful implementation. For fitting GPs in OSP, several themes emerge. First and foremost, the aggregate RMC is a sequence of tasks, indexed by the time steps k. These tasks are recursive; the sub-problem at step k is linked to the previous simulators/emulators at steps $\ell > k$. Therefore, approximation errors will back-propagate. Second, the "error term" in $y(\mathbf{x}) = f(\mathbf{x}) + \epsilon(\mathbf{x})$ is coming from the pathwise reward/value simulator. This stochasticity is deeply embedded in all the pathwise simulators and the dependence across time-steps, making the distribution of $\epsilon(\cdot)$ statistically complicated and non-Gaussian. Hence, correct modeling of ϵ is critical for maximizing performance. For example, in the common scenario where the reward has a fixed lower bound like $G(k, \mathbf{x}) \geq 0$, $y(\mathbf{x})$ and consequently also $\epsilon(\mathbf{x})$ has a mixed distribution with a point mass. Consequently, homoskedastic Gaussian noise is a poor choice for RMC. Indeed, the variance in pathwise rewards is highly state-dependent. For example, deep out-of-the-money the conditional variance of the payoff is negligible, whereas it is very high deep in-the-money.

Third, RMC is often applied in large throughput settings with many thousands of simulations $N \gg 10^4$. Due to the cubic computational complexity of GPs on N, this is not feasible; a practical GP library cannot directly handle a covariance matrix with more than a few thousand dimensions. Replicated designs (cf. Sect. 3.1) offer one way out of this challenge; other solutions could be local GPs or sparse GPs mentioned in Chap. 3. At the same time, given a good training set, GP emulators can provide highly accurate fits even with just a few dozen (well-placed) training $\mathbf{x}^n$. A partial solution to account for this is to combine a replicated design with Stochastic

Kriging (SK) [7] which estimates $\sigma^2(\mathbf{x})$ empirically via the classical MC variance estimator based on the batch of the $\check{N}$ pathwise timing values originating at $\mathbf{x}^n$ (see Eq. (4.5)). SK does require a large batch size $\check{N} \gg 20$ to be reliable.

In multidimensional settings, one usually utilizes a separable kernel, such as the anisotropic SE

$$k(\mathbf{x}, \mathbf{x}') = \eta^2 \prod_{j=1}^{d} \exp\left(-\frac{|x_j - x'_j|^2}{2\ell_j^2}\right)$$

which includes an automatic relevance determination to capture the potentially different scales in different coordinates. Note that one can use an isotropic kernel (2.5), restricting to $\ell_j \equiv \ell$ for all k, in which case all components of $\mathbf{X}_t$ are treated symmetrically (which is unrealistic but may be applicable when all components of $\mathbf{X}$ have identical dynamics).

5.1.2   Training Designs

We refer to [82, 111, 113] for a discussion of experimental design strategies to determine the training domains of the RMC regression emulators. With a slight abuse of notation, we denote by $\mathcal{D}_k$ the collection of training inputs for $\widehat{T}(k, \cdot)$.

A starting observation is that the surrogate $\widehat{T}(k, \mathbf{x})$ is only needed for the in-the-money region

$$\mathcal{X}_{in} \stackrel{\triangle}{=} \{\mathbf{x} : G(k, \mathbf{x}) > 0\};$$

when the immediate payoff is zero, $G(k, \mathbf{x}) = 0$, it is clear that continuing is better and one can immediately set $\widehat{A}_k(\mathbf{x}) = 1$. Thus, without any loss of generality one may restrict $\mathcal{D}_k$ to be entirely in $\mathcal{X}_{in}$. A related fact is that accuracy of $\widehat{T}$ is directly controlled by the location and density of the training inputs. Therefore, accuracy would be highest in regions where there is a high density of $\mathbf{x}$'s; conversely low accuracy would be expected in regions where there are few training inputs, or none at all. In particular, GPs are generally not good extrapolators, so the shape of each $\mathcal{D}_k$ should cover (with some margin) the region of interest. At the same time, financially, $\widehat{T}$ must be accurate where it matters to make correct decisions. Since decisions are made according to the mappings $A_k(\mathbf{X}_k)$, the frequency of needing to decide whether to stop at $(k, \mathbf{x})$ is controlled by how often that state is visited on *forward, controlled paths*. In turn, this is determined by the conditional density of $\mathbf{X}_k | \mathbf{X}_0$ given the initial condition $\mathbf{X}_0$ and the stopping rule (i.e. the joint density of $(\tau, \mathbf{X}_\tau)$). This creates a complex structure regarding which parts of the space-time pairs matter the most.

Early on (suggested as far back as [107] and [28]), a fully *randomized simulation design* was used: generating a collection of N sample paths $\{(\mathbf{X}_s^{(1)}), \ldots, (\mathbf{X}_s^{(N)})\}$ emanating from a given initial condition X_0 and then populating all $\mathcal{D}_k$;s based on $\mathbf{x}^n = \mathbf{X}_k^{(n)}, n = 1, \ldots, N$. This approach reflects the classical dynamic programming idea of backward propagation (in analogue to binomial trees). It captures the full conditional (non-stopped) density of each $\mathbf{X}_k|\mathbf{X}_0$ which is a popular way to define the region of interest, and moreover automatically addresses the issue of handling the a priori unbounded input space $\mathcal{X}$ (such as $\mathbb{R}_+$ for the Bermudan Call). However, in many instances, this strategy is highly inefficient. Consider, for example an out-of-the-money contract, meaning that $G(0, \mathbf{X}_0) = 0$. In that case, most of the forward paths $\mathbf{X}_k|\mathbf{X}_0$ are likely to be out-of-the-money for most of time steps. As a result $\{\mathbf{X}_k^{(n)}\}_{n=1}^N$ would be highly mis-aligned with $\mathcal{X}_{in}$, making the fit $\ddot{T}(\cdot)$ inefficient. In fact, this issue runs deeper, as the forward paths $\mathbf{X}_k^{(n)}$ will naturally explore the at-the-money region more than in-the-money one. However, from a decision-making perspective, it is easy to ascertain that $\widehat{T}(k, \mathbf{x}) > 0$ close to at-the-money (typically the timing value is *maximized* at-the-money and hence is highly positive there), and much more difficult to decide when exactly to stop in-the-money. Thus, it makes sense to explore the most around the stopping boundary, which is almost the opposite of what tends to happen with the above design strategy.

Customized Training Designs While the size $N_k = |\mathcal{D}_k|$ of the training design obviously plays a major role in how accurate the approximation will end up, the *geometry* of $\mathcal{D}_k$ also has a significant impact. To represent the design geometry we consider the *training densities* $p_k(\cdot)$ that capture the distribution of $\mathbf{x} \in \mathcal{D}_k$ relative to the overall $\mathcal{X}$. In the described randomized simulation design, samples are directly generated from p_k: each $\mathcal{D}_k$ is populated by sampling from the conditional density of $\mathbf{X}_k|\mathbf{X}_0$. Thus in the common case where the dynamics of $\mathbf{X}$ are Geometric Brownian Motion, the p_k are log-normal and concentrate on regions reachable by $\mathbf{X}_k$ starting from $\mathbf{X}_0$. *Uniform* training densities, $p_k \equiv \mathrm{Unif}_{\tilde{\mathcal{X}}}$ for a given bounded input domain $\tilde{\mathcal{X}}$, are also common and correspond to space-filling simulation designs. These capture the intuition of learning through exploration, i.e. sampling a diverse set of $\mathbf{x}^n$'s in order to observe the correspondingly diverse y's. However, this approach requires the user to specify the supporting bounding region $\tilde{\mathcal{X}}$, something not needed with randomized simulation design.

Deterministic construction of $\mathcal{D}_k$ is an alternative, such as employing lattice or Quasi Monte Carlo (QMC) low-discrepancy sequences, for example Sobol, Halton, and Faure sequences. These minimize discrepancy of samples across any sample size N. For applications requiring randomness, scrambled QMC sequences enable randomized designs by facilitating random steps through the sequence. Latin hypercube sampling (LHS) [128] presents another flexible approach, allowing for random space-filling designs. LHS can also be easily combined with an acceptance-rejection step to ensure efficient exploration of the input space. As a random method, this significantly improves over basic strategies like i.i.d. Uniform sampling, especially in high dimensions where it prevents the clustering and gaping of inputs, thus enhancing learning quality by ensuring an equidistributed subset across the input space.

The above methods can also be catered to the OSP instance at hand. For example, [75] uses a Halton quasi-random sequence combined with the inverse cdf of a standard normal to produce a quasi-Monte Carlo sampling for the $\mathcal{N}(0, 1)$ component of the log-normal draws of $\mathbf{X}_k|\mathbf{X}_0$. This maintains a quasi-log-normal target training density p_k while effectively covering the region of interest. This technique complements the space-filling simulation designs, aiming to capture a diverse array of inputs through varied sampling within the bounding region $\tilde{X}$, thus facilitating an exhaustive exploration without specifying a supporting bounding region, as is necessary with uniform training densities.

Additional degree of freedom is provided by the fact that $\mathcal{D}_k$ is generated separately for each time step k. As such, one can straightforwardly also vary the design size N_k. One reason why varying N_k is useful is to reflect the growing volume of the region of interest as k increases. Indeed, learning $A_k(\cdot)$ over a larger domain necessarily requires more simulation effort. Moreover, backpropagation of approximation errors implies that the algorithm is more sensitive to errors in the middle/end time-steps than in the early ones.

Used in tandem with the methods above is the *batched design* approach. Here, $\mathcal{D}_k$ consists of $n_k \ll N_k$ unique input locations, with a_i replications at $\mathbf{x}^i$. The GP for $\widehat{T}(k, \cdot)$ is then built based on the pre-averaged reduced dataset $(\mathbf{x}^{1:n_k}, \bar{y}^{1:n_k})$, see the end of Sect. 3.1 for details. This approach significantly reduces computation time while also giving a natural way to handle heteroskedastic noise. Replication is a must to be able to tractably implement GP regression in a moderately-high dimensional problem that requires $N_k \gg 10^3$.

5.1.3 Illustration: Bermudan Options

We illustrate with two examples of Bermudan options, the first being for a univariate Put, and the second a basket average Put on two assets. In both instances the inputs are the underlying asset prices. For the first case study, the payoff is $G(t, x) = e^{-rt}(\mathcal{K} - x)_+$ with underlying dynamic given by Geometric Brownian Motion (GBM)

$$dX_t = (r - \delta)X_t \, dt + \sigma X_t \, dW_t, \qquad X_0 = x_0, \tag{5.11}$$

with model parameters r, δ, σ, x_0. In discrete-time, $\mathbf{X}_k$ is conditionally log-normal distribution and can be simulated exactly,

$$\log X_k \sim \mathcal{N}\left(\log(X_0) + (r - \delta - \frac{1}{2}\sigma^2)k\Delta t, \sigma^2 k\Delta t\right).$$

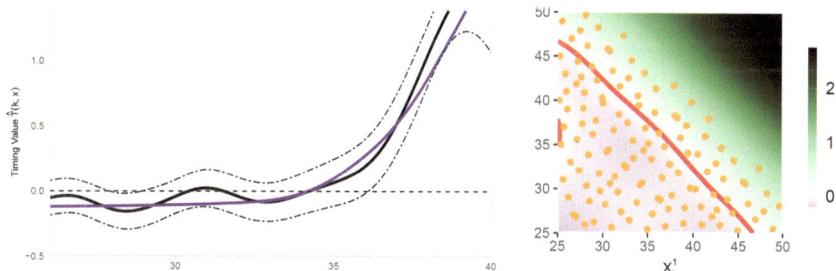

Fig. 5.1 Left: Timing value of a 1D Bermudan Put based on GP emulator (black) and a Smoothing Spline emulator (purple) at $k = 10$. Also shown is the dashed 95% band of $\widehat{T}(k, \cdot)$. Right: Timing Value of a 2D Basket Put at $k = 15$ estimated via GP surrogate with SE kernel (2.6). The red contour shows the boundary of the stopping set (bottom-left corner). The colors indicate the value of $\widehat{T}(k, \mathbf{x})$

We consider above with parameters $r = 0.06, \delta = 0, T = 1, \sigma = 0.2$, initial $X_0 = 40$ and at-the-money strike $\mathcal{K} = 40$. Exercising the Put is possible $K = 25$ times before expiration yielding a spacing of $T/K = \Delta t = 0.04$. We build a GP surrogate with Matérn-5/2 kernel (2.11) using fixed hyperparameters $\eta = 1, \ell_{\text{len}} = 4$, and a constant prior mean $\mu(\mathbf{x}) = \beta_0$. For the simulation design, we only need the GP for in-the-money locations $X_{in} = \{x : x \le 40\}$; moreover X_k cannot be too small given $X_0 = 40$. Accordingly, we pick a uniform grid of $n = 25$ locations $\mathbf{x}^i \in \{16, 17, \ldots, 40\}$ for all $\mathcal{D}_k$, with $a_i = 200$ replications for all i, yielding a total design size of $N = |\mathcal{D}_k| = 200 \cdot 25 = 5000$. The replications are treated using the SK approach (see (3.7)), pre-averaging the replicated outputs before training the GP. Note that this means that effectively the GP is trained on just 25 outputs which were heavily de-noised first. Figure 5.1 visualizes the fitted timing value from one time step. To do so, we predict $\widehat{T}(k, x)$ based on the fitted GP model at $k = 10$ (corresponding to the business time $t_k = 0.4 = 10\Delta t$) over a collection of test locations x_*. In the Figure this is done for two different solvers (GP and Spline), moreover we also display the 95% credible intervals of the GP emulator for $\widehat{T}(k, \cdot)$. We emphasize that in the OSP application, the quality of the solution is based on the implied stopping rule $\widehat{A}_k(\cdot) = 1_{\widehat{T}(k,\cdot)>0}$, so what matters is only the *sign* of the plotted $\widehat{T}(k, \cdot)$. The credible band corresponding to $\widehat{T}(k, \cdot)$ is a useful diagnostic to assess the GP-based confidence $\widehat{A}_k$. In Fig. 5.1 the displayed uncertainty band shows that the GP emulator has low confidence about the right action to take for nearly all $x \le 36$, since zero is inside the 95% band. Note that the training design here is the same for different time steps; in general, we need a more voluminous $\mathcal{D}_k$ for larger k to reflect the intrinsic spreading of X_k as k increases.

The right panel of Fig. 5.1 considers a two-dimensional OSP where the two assets $X^{(1)}, X^{(2)}$ are assumed to be uncorrelated with identical Geometric Brownian motion dynamics as in (5.11) and the payoff function the basket average Put: $G(t, \mathbf{x}) = e^{-rt}(\mathcal{K} - (x_1 + x_2)/2)_+, \mathbf{x} \in \mathbb{R}_+^2$. The strike is $\mathcal{K} = 40$ with at-the-money initial condition $\mathbf{X}_0 = (40, 40)$. Hence, an initial approximation of the region of interest is $\{(x_1, x_2) : x_1 + x_2 \leq 80, |x_i - 40| \leq 20\}$, leading us to pick a space-filling, triangular-shaped training design. Specifically we employ a GP surrogate with SE kernel (2.6) trained at $n = 150$ sites replicated with batches of $a_i = 100$ each for a total of $N = 15,000$ simulations. We show an image plot of the emulator $\widehat{T}(k, \mathbf{x})$ at a single time-step k and its zero level set, indicated with the red contours that delineate the exercise boundary. The stopping rule is to stop when inside (bottom-left) the contour and continue otherwise. In this example, because the two coordinates are fully symmetric, the stopping rule is expected to be symmetric in x_1, x_2 (i.e. around the $x_1 = x_2$ 45-degree line). This is one instance where, given the identical dynamics of $X^{(1)}$ and $X^{(2)}$, an isotropic (2.5) kernel is appropriate.

Simulating SDEs

The given presentation is in discrete time. In most financial contexts, one starts with a continuous-time process $(\mathbf{X}_t)$ which is then discretized before solving the OSP. Commonly, $(\mathbf{X}_t)$ is defined via a (multi-dimensional) SDE, and exact simulation of $\mathbf{X}_{t_k}|\mathbf{X}_0$ is not possible. In that case, one first constructs an approximate numerical scheme, for example using Euler-Maruyama methods, to draw samples of $\mathbf{X}_{t_{k+1}}|\mathbf{X}_{t_k}$—the latter using a time-step δt not to be confused with the Δt that governs exercise opportunities and indexes the timing values and value functions above. In situations where the dynamics of $(\mathbf{X}_t)$ are complex, a very small δt might be required making such simulations relatively expensive which should be factored when optimizing the computational overhead across constructing $\mathcal{D}_k$'s, fitting $\widehat{T}(k, \cdot)$'s and evaluating y_k^n's.

5.2 Active Learning and Adaptive Batching

As discussed in Ludkovski [111] and suggested from the above examples, it is most efficient to select inputs around the exercise boundary, i.e. the region where the correct decision rule is most difficult to ascertain. Of course, the exercise boundary is unknown a priori, motivating a sequential construction $\mathcal{D}_k$ now indexed by m, gradually adding training samples $\mathbf{x}^{(m)}$. The goal of such adaptive $\mathcal{D}_k^{(m)}$ is to improve simulation efficiency through targeted placement of $\mathbf{x}^{(m)}(k)$'s to maximize the learning of the timing value. Such *active learning* methods are common in machine learning applications.

The primary aim of adaptive $\mathcal{D}_k^{(m)}$ is to improve upon the space-filling designs that will often have many simulation sites far from the stopping boundary (e.g. deep in-the-money) and which are therefore not informative about the optimal stopping rule. At the same time, one must consider the exploitation-exploration trade-off, i.e. not to neglect learning $\widehat{T}$ sufficiently in other parts of the state space. Ludkovski

[111] implements active learning of the exercise region through greedily optimizing an *acquisition function* $\mathbf{x} \mapsto I^{(m)}(\mathbf{x})$ that is a proxy for the information gain for the respective input location $\mathbf{x}$. The selection of $\mathbf{x}^{(m)}$ is done one-by-one by maximizing $I^{(m)}(\cdot)$. Such acquisition functions rely on the posterior uncertainty of $\widehat{T}^{(m)}(k, \cdot)$ provided by the GP surrogate. This method echoes active learning heuristics in Bayesian Optimisation: it introduces an auxiliary step that tries to find the most *informative* training inputs in order to converge to the true A_k rule as quickly as possible. To this end, rather than space-fill training inputs, they are placed in areas which matter the most. In turn, given an updated $\mathcal{D}_k^{(m+1)}$, one updates $\widehat{T}^{(m+1)}(k, \cdot)$ using the logic in Sect. 3.5.

Active learning is initialized with an initial (typically space-filling) $\mathcal{D}_k^{(0)}$ and then gradually augmented until a high confidence in the $\widehat{A}_k$ is achieved. The required logic of evaluating $I^{(m)}$ and optimizing for $\mathbf{x}^{(m+1)}$ introduces additional overhead that must be carefully accounted for in implementation. Indeed, sequential design is highly efficient in terms of keeping N low, but is rather slow due to shifting the work from running simulations to updating the GP. The gains from sequential design are biggest in models where simulation is expensive (e.g. where very small simulation steps are needed to generate $\mathbf{X}_{k+1}|\mathbf{X}_k$), or where high replication is warranted.

The left panel of Fig. 5.2 illustrates a sequential design constructed using the straddle Maximum Contour Uncertainty sMCU acquisition heuristic [111]. This rule uses

$$I^{(m)}(\mathbf{x}) = -|\widehat{T}^{(m)}(k, \mathbf{x})| + \gamma^{(m)} s^{(m)}(\mathbf{x}),$$

where $s^{(m)}$ is the posterior standard deviation of the GP surrogate $\widehat{T}^{(m)}$ and $\gamma^{(m)}$ is a user-defined sequence, taken to be the constant $\gamma^{(m)} \equiv 1.96$ in this example. Hence, $I^{(m)}$ is largest when $\widehat{T}^{(m)}(k, \cdot)$ is close to zero (around the stopping boundary) and/or when the posterior standard deviation $s^{(m)}(\mathbf{x})$ is large (in sparsely

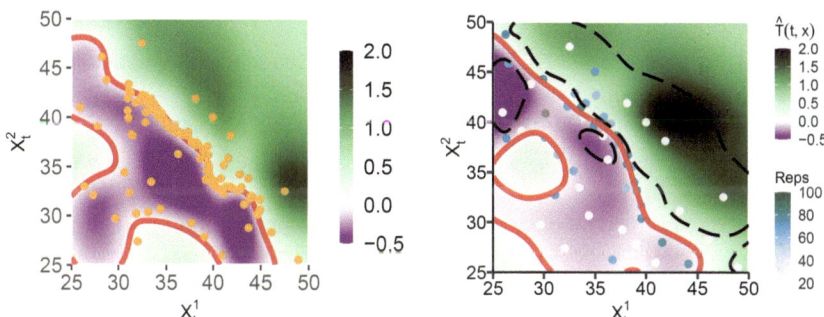

Fig. 5.2 Left: Sequential design with sMCU. Right: Adaptive batching with ADSA. Both designs are at $k = 15$ ($t = 0.6$) for the 2D Basket Put example. Replication counts $r^n(k)$ are input-dependent and color coded in grayscale. The surface plots are color-coded according to $\widehat{T}(k, \cdot)$. Taken from [113]

sampled regions). The weight $\gamma^{(m)}$ balances this exploration-exploitation trade-off and resembles the Upper Confidence Bound (UCB) rule in Bayesian Optimization.

We start with $m_0 = 30$ space-filling design locations to populate $\mathcal{D}_k^{(m_0)}$, and augment with additional 90 inputs with $a = 25$ replications each, for a total $|\mathcal{D}_k| = 120 \cdot 25 = 3000$ simulations. The resulting training design $\mathcal{D}_k$ at time step $k = 15$ (i.e. $t = 0.6$) has a total of $n = 120$ unique training inputs. Compared to the space-filling designs like in the right panel of Fig. 5.1, active learning yields a highly non-uniform placement of $\mathbf{x}^{1:N} \in \mathcal{X}$, placing them *around the stopping boundary*, in other words the zero-contour of the timing value. See [111] for a detailed analysis of sequential designs in RMC, including the trade-off between number of unique inputs n and replication counts a_i, (keeping $N = \sum_{i=1}^n a_i$ fixed).

Conceptually, active learning is intended to favor locations close to the exercise boundary where the correct decision rule is hardest to resolve. As a result, sequential designs will increasingly concentrate, i.e. the added $\mathbf{x}^{(m)}$'s cluster as m grows. *Adaptive batching* takes advantage of this by gradually increasing the replication level, in effect replacing clusters of similar but slightly varying $\mathbf{x}$'s with a single replicated input. This allows to reduce the number of unique inputs n and speeds up the construction of the sequential design. In an adaptively batched design, the constant a_i associated with each unique $\mathbf{x}_i$ in (3.4) is replaced with input-dependent sequential replication counts $a_i^{(m)}$:

$$\mathcal{D}_k^{(m)} = \Big\{ \underbrace{\mathbf{x}_1, \mathbf{x}_1, \ldots, \mathbf{x}_1}_{a_1^{(m)} \text{ times}}, \underbrace{\mathbf{x}_2, \ldots, \mathbf{x}_2}_{a_2^{(m)} \text{ times}}, \ldots, \underbrace{\mathbf{x}_n, \mathbf{x}_n, \ldots, \mathbf{x}_n}_{a_n^{(m)} \text{ times}} \Big\}, \qquad (5.12)$$

where the algorithm now specifies both the unique inputs (meaning n may increase with m) and the respective $a_1^{(m)}, a_2^{(m)}, \ldots$. This idea was explored in detail in Lyu and Ludkovski [120] that proposed several strategies to construct $a_i^{(m)}$ sequentially. In essence, more replication is warranted closer to the stopping boundary, which can be proxied by either $\widehat{T}^{(m)}(k, \mathbf{x}^{(m+1)})$ or more simply by m—replicate more as the number of unique inputs grows. The right panel of Fig. 5.2 shows the output from the batching heuristic termed Adaptive Design with Sequential Allocation (ADSA), [120] at one intermediate step $k = 15$ which yields a design $\mathcal{D}_{15}$ with $n = 65$ unique inputs and with replication counts ranging from 10 and up to $a_i^{(m)} = 79$. Note that the geometry of $|\mathcal{D}_k|$ is very similar to that in Fig. 5.2 left, but n is almost half as small. Thus, adaptive batching allows to reduce the number of sequential design rounds and the associated computational overhead, running multiple times faster.

Further Reading

We refer to [113] for a comprehensive summary of GPs for RMC, see earlier [82, 111, 120]. A broader treatment of computational optimal stopping is in the monograph [14]. Many simulation-based algorithms, including several flavors of GP-based approaches, are in the R library mlOSP.

This Chapter includes an online supplementary R notebook illustrating construction of GP surrogates for one- and two-dimensional Bermudan option valuation.

Chapter 6
Non-Parametric Modeling of Financial Structures

This chapter investigates a larger span of GP models for financial markets, including one-dimensional term structures, two-dimensional volatility surfaces (both implied and local), three-dimensional swaption cubes, and valuation of variable annuities. Additionally, this discourse extends into the realms of mortality modeling and actuarial mathematics, illustrating a wider applicability of GP models.

6.1 Modeling Term Structure

A term-structure is a curve which describes the evolution of a financial or economic quantity as a function of time to maturity. Typical examples are the term-structure of risk-free interest-rates, the term-structure of bond yields or credit spreads, and the term-structure of default probabilities. These curves are not directly observed and are inferred based on the prices of financial instruments whose values depend on the curve. The curve construction problem is concerned with transforming a given set of market quotes into a continuum of values representing the full term-structure function. The quality of the produced term-structure estimate directly impacts the quality of the downstream applications that build on it.

Mathematically, in the one-dimensional case, the task is to infer the mapping $\tau \mapsto P(t, \tau)$ or $(t, \tau) \mapsto P(t, \tau)$, where t is current time, $\tau = T - t$ is time to maturity, and $P(t, \tau)$ is a term structure, e.g. the time t price of a default-free zero-coupon bond with time to maturity τ. The term structure is typically inferred from market price(s) $S(t, \tau)$. Continuing with the bond example, denote the fixed coupon rate as c (expressed in percentage of invested nominal) and coupon dates as $\tau_1, \ldots, \tau_K = \tau$, then the (observed) coupon bond price is given by

$$S(t, \tau) = c \sum_{k=1}^{K} (\tau_k - \tau_{k-1}) P^B(t, \tau_k) + P^B(t, \tau_K), \qquad (6.1)$$

where we define $\tau_0 = 0$ and P^B is the (unobserved) zero-coupon bond. The process of switching from a collection of observed prices $S_n(t, \tau_n)$, $n = 1, \ldots, N$ (which may have different coupon dates among them) to the underlying $P^B(t, \cdot)$ is known as *stripping* the bonds. Another example is that of an overnight indexed swap (OIS), where one observes the par swap rates S_n and wishes to infer the OIS discount curve $P(t, \tau)$. This is done from the swap equilibrium relation between the fixed legs (on the left hand side below) and the floating leg present value

$$S_K \sum_{k=1}^{K} (\tau_k - \tau_{k-1}) P(t, \tau_k) = 1 - P(t, \tau). \qquad (6.2)$$

Cousin et al. [35] present more examples about constructing forward curves based on OIS and fixed versus Ibor-floating interest-rate swaps and Credit curves based on Credit Default Swaps spreads. Note that both (6.1) and (6.2) can be represented as the general linear system

$$\mathbf{A} \cdot \mathbf{P} = \mathbf{b}, \qquad (6.3)$$

where $\mathbf{b} = [S_1(t, \tau_1), \ldots, S_N(t, \tau_N)]^\top$, $\mathbf{P} = [P(t, \tau_1), \ldots, P(t, \tau_N)]^\top$, and $\mathbf{A}$ is an $N \times N$ matrix so that (6.1) (or (6.2)) holds. Note that $\mathbf{A}$ may depend on t and the various τ. Curve construction needs to satisfy several conditions to be financially relevant:

Price compatibility: if one uses the constructed curve to value a given set of benchmark instruments, the resulting values should match observed market quotes. If the market quotes are exact, this gives a system of equalities (i.e., linear constraints); more generally assuming bid-ask spreads or the quotes $S(t, \cdot)$ not being fully reliable gives a set of inequalities or alternatively noisy observations.

Monotonicity constraints: in many applications the term-structure must be monotone. For instance, the price of default-free zero-coupon bonds (or risk-free discount factors) is a non-increasing function of time-to-maturity. Similarly, survival functions inferred from credit default swap (CDS) spread term-structures are $[0, 1]$-valued non-increasing functions.

No-arbitrage shape restrictions: in addition, there may be additional constraints (such as convexity or linear bounds) on the curve implied by financial no-arbitrage considerations. For instance, the function $T \to P(t, T)$ may be known to be decreasing with respect to time horizon T and its values may be bounded. This is typically the case when one constructs a curve of discount factors (default-free zero-coupon bond prices) or an implied survival function (survival probabilities of a CDS reference entity), both of which must belong to the interval [0, 1]. Violating these kinds of shape-preserving conditions results in term-structure functions that are typically not arbitrage-free.

Uncertainty quantification: accounting for the uncertainty embedded in the process of curve construction is important to quantify model risk and to capture the unreliability of market quotes. The construction of marginal distributions or term-structure functions themselves matters for risk management, whether it may concern discount curves, zero-coupon curves, swap basis curves, bond term structures or CDS-implied survival curves.

In all, one seeks a non-parametric and robust curve construction method that can explain complex term-structure shapes and offers a good trade-off between flexibility and smoothness of the curve. Historically, term-structure curves were constructed via parametric models. These reduce inference to learning a finite set of coefficients, typically estimated using least-squares regression in accordance with the price compatibility requirement. For example, the 5-parameter Nelson-Siegel parametric family (6.7) below is commonly used to fit term structures of interest rates. However, parametric or deterministic methods yield a single point forecast, with no associated uncertainty.

In line with the theme of this book, we present a GP model $P \sim \mathcal{GP}(\mu, k)$ based upon (6.3) and incorporating pricing errors $(\epsilon_n)_{n=1}^N$: $\mathbf{A} \cdot \mathbf{P} + \boldsymbol{\epsilon} = \mathbf{b}$. Measurement errors arise due to market imperfections, such as lack of liquidity and data errors, and have important consequences on the dynamics of yield curves [101]. For the discussion herein, they simply capture deviations of observed prices $(S_n)_{n=1}^N$ from their fundamental values. The resulting GP posterior mean and variance of $P(\mathbf{x}_*)$ are

$$m(\mathbf{x}_*) \overset{\triangle}{=} \mu(\mathbf{x}_*) + (\mathbf{A}k(\mathbf{X}, \mathbf{x}_*))^\top (\mathbf{A}\mathbf{K}\mathbf{A}^\top + \boldsymbol{\Sigma})^{-1}(\mathbf{b} - \mathbf{A}\mu(\mathbf{X})) \qquad (6.4)$$

$$s^2(\mathbf{x}_*) \overset{\triangle}{=} k(\mathbf{X}, \mathbf{x}_*) - (\mathbf{A}k(\mathbf{X}, \mathbf{x}_*))^\top (\mathbf{A}\mathbf{K}\mathbf{A}^\top + \boldsymbol{\Sigma})^{-1}(\mathbf{A}k(\mathbf{X}, \mathbf{x}_*)). \qquad (6.5)$$

For the observation noise, one possibility is to define $\boldsymbol{\Sigma}$ as a diagonal matrix where each positive term is proportional to the squared difference between ask and bid quotes. Additionally, monotonicity constraints can be imposed as in Sect. 4.4.

Filipovic et al. [59] use kernel ridge regression to infer yield curves from prices of coupon bonds, using a weighted norm that penalizes both the first and second derivative of the discount curve. This approach allows the kernel k to be inferred through an associated RKHS norm. The authors consider a version of KRR (1.31) given by

$$\hat{P} = \operatorname{argmin}_{P \in \mathcal{H}_{\alpha,\delta}} \frac{1}{N} \sum_{n=1}^N w_n (b_n - A_n P)^2 + \lambda \|P\|_{\alpha,\beta}^2,$$

using weight w_n for the n-th coupon bond and a penalty norm of the form $\| f \|_{\alpha, \delta}^2 \triangleq$
$\int_0^\infty \left(\delta f'(x)^2 + (1 - \delta) f''(x)^2 \right) e^{\alpha x} dx$ where $\alpha \geq 0$ and $\delta \in (0, 1)$. One of the
motivations for this norm is because all arbitrage-free discount curves must be twice
differentiable [59, Theorem 1]. Moreover, that reference emphasizes the preference
for the prior mean $\mu(x) = 1$ that corresponds to zero interest yields. The KRR
perspective gives the following bespoke kernel for $\alpha > 0$:

$$k_{\alpha, \delta}(x, x') = -C_1 \left(1 - e^{-\ell_2 x} - e^{-\ell_2 x'} \right) + C_2 (1 - e^{-\alpha(x \wedge x')}) \tag{6.6}$$

$$+ C_3 \left(\frac{\ell_1^2}{\ell_2^2} e^{-\ell_2(x + x')} - e^{-\ell_1(x \wedge x') - \ell_2(x \vee x')} \right),$$

where the constants $C_1, C_2, C_3, \ell_1, \ell_2$ are all explicitly specified in terms of α, δ.
The pricing error weights w_n are taken to be inversely proportional to the squared
duration of the coupon bond $w_n \propto (D_n \cdot b_n)^{-2}$, making the overall observation
matrix Σ in (6.4) be of the form $\Sigma_{nn} = \lambda / w_n$.

For the KRR training step, [59] estimate the respective smoothness penalty λ
and the norm parameters α, δ by cross-validation based on minimizing the out-of-
sample pricing error from the fitted discount curve. The authors find that $\alpha \simeq 0.05$
and $\delta \in [0, 0.001]$. After fitting the point estimate of the yield curve, uncertainty
quantification is obtained ad hoc by using the KRR/GP link (1.32) to obtain posterior
confidence intervals for the estimated discount curve, yields, and implied fixed
income security prices. One application is yield curve extrapolation; the advantage
of the method is that the only exogenous choice parameter needed has a clear
economic interpretation as the infinite maturity yield. As expected, extrapolation
based on GPs generates a wide cone of uncertainty. See [25] for a further application
of KRR to Swiss bond data.

Remark 6.1 For no exponential weighing, $\alpha = 0$, (6.6) simplifies to

$$k(x, x') = \frac{1}{\delta}(x \wedge x') + \frac{1}{2 \delta \rho} \left(e^{-\rho(x + x')} - e^{-\rho(x \wedge x') - \rho(x \vee x')} \right)$$

with $\rho = \sqrt{\delta / (1 - \delta)}$ and for no convexity-tension penalty $\delta = 0$ to

$$k(x, x') = -\frac{x \wedge x'}{\alpha^2} e^{-\alpha(x \wedge x')} + \frac{2}{\alpha^3}(1 - e^{-\alpha(x \wedge x')}) - \frac{x \wedge x'}{\alpha^2} e^{-\alpha(x \vee x')}.$$

Note how the above is a linear combination of Exponential (Matérn-1/2) (2.9) and
compound Exponential-Minimum kernels.

6.1.1 Kriging of Commodity Curves

Commodity markets usually quote futures contracts prices at a discrete set of maturity dates. The futures contracts have different delivery periods, and there are also traded spread contracts. For instance, natural gas is quoted with delivery periods of one, three, six and twelve months, along with spreads between consecutive one-month or three-month contracts. The goal becomes to model and then calibrate the term-structure $T \to f_t(T)$ of futures prices indexed by the shortest delivery period.

Due to the physical nature of the underlying assets, commodity futures feature *periodic* structures driven by seasonal effects and fluctuating levels of supply and demand. For instance, in US and Europe there is more natural gas demand (for heating) in winter than in summer. In order to correctly describe such seasonalities, additional machinery, for example via trigonometric functions with different time scales, is common [110]. In order to learn these, sufficient resolution is required from observed quotes. For example, natural gas prices display an annual pattern. Yet, futures price for contracts with maturities more than 2–3 years are usually not quoted except for quarterly maturity dates, or are quoted with wide (3+ months) delivery periods. Thus, the term-structure model faces the problem of extrapolating the seasonal pattern.

In sum, a GP surrogate for commodity curves must be able to (i) model different complex shapes, (ii) deal with seasonalities, (iii) consider bid-ask spreads, (iv) bootstrap quotes from the prices of overlapping linear products. To our knowledge, the first article to propose GPs in the context of commodities is Benth [18], using a GP $f(\cdot)$ to model the forward curves of crude oil and natural gas. He proposed to take a parametric prior mean $\mu(\cdot)$, namely to use the Nelson-Siegel model with

$$\mu_{NS}(T) \triangleq \alpha_0 + \alpha_1 e^{-\beta T} + \alpha_2 \beta t e^{-\beta T}. \tag{6.7}$$

The coefficients $\alpha_0, \alpha_1, \alpha_2, \beta$ were pre-estimated separately (rather than within a universal kriging (1.29) setup) using least squares regression. For natural gas forward curves, [18] suggested a further pre-processing step to remove the annual seasonality, and then fit the GP on the de-trended residuals. The author fitted the GP using a semivariogram based on historical correlation of the forwards. Specifically, he used the variogram to estimate the coefficients of an exponential Matérn-1/2 kernel. The estimated lengthscale was $\ell_{len} \simeq 0.02$ in daily units, so that futures for two consecutive months have a correlation of about $\text{corr}(f(T), f(T + 30)) = e^{-0.02 \cdot 30} = 55\%$.

Extending [18], Maran and Pallavicini [125] construct a finite-dimensional GP f for the market quotes of futures contracts across both maturities and different delivery periods and also take into account the respective bid-ask spreads. Consider a grid of maturities $T_1, \ldots, T_r$. Given ask quotes $q_n^a, n = 1, \ldots, N$ and corresponding bid quotes $(q_n^b)_{n=1}^N$ of a forward contract with delivery period $[T_n, T_n']$, the no-arbitrage constraints are similar to (6.3):

$$q_n^b \leq \sum_{i=1}^r A_{i,n} f(T_i) \leq q_n^a. \tag{6.8}$$

For instance, if the n-th futures contract has a delivery period of three months starting at T_3 and terminating in T_6, the coefficients are $A_{i,n} = (T_i - T_{i-1})/(T_6 - T_4)$ for $i \in \{4, 5, 6\}$ and zero for the other values of i. Letting q_n^{mid} denote the midprice for the n-th maturity, the market mispricing is expressed as the errors $\epsilon_n = \sum_{i=1}^r A_{i,n} f(T_i) - q_n^{mid}$. Maran and Pallavicini [125] use Gaussian observation noise $\epsilon_n \sim \mathcal{N}(0, \sigma_n^2)$ with variance $\sigma_n^2 = (q_n^a - q_n^b)^2/4$.

To guarantee the constraints in (6.8), [125] employ the construction in Sect. 4.4, using the linear interpolator $\bar{f}(T_i) = \sum_{k=1}^{N_T} \xi_k \cdot \left(1 - \frac{T_i - t_k}{t_{k+1} - t_k}\right)_+$ where (ξ_k) is a finite-dimensional GP and the inducing points t_k are on a regular grid $\{t_k : k = 1, \ldots, N_T\}$ with spacing $\Delta t = t_{k+1} - t_k$. Observe that the last terms in the sum are the univariate hat functions $\phi_k(t) = \max(1 - |t - t_k|/\Delta t, 0)$, compare to (4.16), and link to a linear B-spline model with a Bayesian MVN prior on the "coefficient" ξ's.

Let $\boldsymbol{\Phi}$ be the $r \times N_T$ matrix with entries $\Phi_{i,k} \overset{\triangle}{=} \phi_k(T_i) = \left(1 - \frac{T_i - t_k}{\Delta t}\right)_+$. Combining with the observations according to (6.8), gives the covariance matrix of (ξ_k) as

$$\mathbf{C} = \boldsymbol{\Sigma} + \mathbf{A}^\top \boldsymbol{\Phi} \mathbf{K} \boldsymbol{\Phi}^\top \mathbf{A}.$$

To enforce annual seasonality, the log-likelihood is augmented with a penalty term regarding the convexity of the futures curve at the same day across different years $\sum_k \partial_T^2 f(t_k) - \partial_T^2 f(t_k - 365)$, where the gradient is approximated via

$$\partial_T^2 f(t_k) \simeq \frac{1}{\Delta t}(\xi_{k-1} - 2\xi_k + \xi_{k+1}). \tag{6.9}$$

Given a penalty parameter λ and letting $\mathbf{q}$ be the N-vector of q_n^{mid}'s, the resulting log-likelihood is $\ell(\boldsymbol{\theta}) \overset{\triangle}{=} \log |\mathbf{C}| + \mathbf{q}^\top \mathbf{C}^{-1} \mathbf{q} + \lambda \sum_{k=2}^{N_T-1}(\xi_{k-1} - 2\xi_k + \xi_{k+1})$. Maran and Pallavicini [125] use maximum likelihood estimation for the hyperparameters of k, modeled as a univariate SE kernel.

To illustrate practical aspects of term-structure construction, we build a GP model for natural gas (NG) forward curve based on NYMEX data. The respective forward contracts $F(T_n)$ (styled NGXYY on NYMEX) reference Henry Hub and have monthly tenors T_n ranging from 1 to 144 months (12 years out). Beyond the first three years the liquidity diminishes rapidly so that there are almost no trading volume outside of the benchmark October and March contracts. Consequently, though the full 144 forwards are quoted daily, the actual market data is much sparser and the other quotes are effectively interpolated by data providers, offering a good

Fig. 6.1 Fitted GP model $T \mapsto F(T)$ to NYMEX natural gas forward curve from 9/10/2024

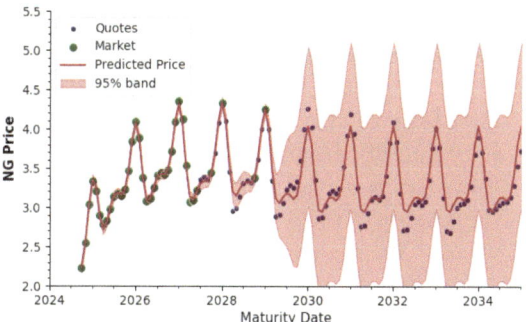

use case for a GP emulator that can quantify the respective quote uncertainty and provide a bespoke model-based interpolation. Since natural gas exhibits a strong annual seasonality, we use an additive kernel consisting of a periodic term like in (2.16) and a regular term like SE (2.5) or rational quadratic (2.12).

Accordingly, we proceed to fit a compound univariate GP model with $k(T_1, T_2) = \eta_{per}^2 k_{per}(T_1, T_2; \ell_{\text{len},1}) + \eta_{se}^2 k_{SE}(T_1, T_2; \ell_{\text{len},2})$ (T measured here in years) to a representative end-of-day set of NG forward quotes. The prior mean is constant $\mu \equiv \mu_0$, set to the average observed liquid quote. On that day there are 37 liquid contracts (chosen to be those with daily volume of at least 3 contracts), indicated by the dark green circles in Fig. 6.1; the respective "Last" trade prices are used as the training inputs $P_t(T_n), n = 1, \ldots, 37$ with t fixed. Note that the first 30 tenors are all liquid and after that liquidity is irregular, with the longest training maturity being March 2029, $T_N = 4.5$ years. The remaining quotes (smaller blue circles) are held out for out-of-sample validation. Since the annual periodicity is assumed, we do not estimate the respective period length p, fixing it to be 1. The kernel hyperparameters are estimated using MLE based on a stochastic gradient descent optimizer.

The resulting fit and 95% predictive bands are displayed in Fig. 6.1. As expected, the uncertainty grows for longer maturities where the training data is very sparse, and we observe the characteristic football-shape around months where quotes *are* available. The obtained coefficients are $\eta_{se} = 0.498, \ell_{\text{len},2} = 0.506, \eta_{per} = 0.569, \ell_{\text{len},1} = 1.312$. We observe that there is a humped shape to the forward curve after removing the annual pattern and that the magnitudes of the two terms are comparable. Moreover, the annual pattern is far from sinusoidal, which is one advantage of working with periodic GPs compared to manually set trigonometric sums.

Moving from Linear Systems to Integrals

In many markets, including natural gas case above, observed forward prices are averages across a delivery period $F(T, T + L) = \frac{1}{L} \int_T^{T+L} f(u)du$ where L is typically 1 or 3 months and the goal is to model the (instantaneous) futures curve $T \mapsto f(T)$. To build a corresponding GP entails extending (6.3) from a linear system to an integral one. Using the linearity of the Gaussian distribution,

$$\mathrm{cov}(F(T, T + L), f(T')) = \frac{1}{L} \int_T^{T+L} \mathrm{cov}(f(u), f(T'))du.$$

The above integral can be evaluated analytically for the exponential k_{M12} kernel. A similar computation pertains to the covariance of $F(T, T + L)$ and $F(T', T' + L)$ in terms of the underlying kernel k of f. Given a collection of (equally spaced) futures with delivery periods $([T_n, T_n + L])_{n=1}^N$, the GP posterior mean is then [129, Proposition 1]

$$m_*(T_*) = \mu(T_*) + \mathbf{c}(T_*)^\top \mathbf{C}^{-1}\mathbf{y}, \tag{6.10}$$

where $\mathbf{c}(T_*) = [c_1(T_*), \ldots, c_N(T_*)]^\top$, $c_n(T_*) = \mathrm{cov}(f(T_*), F(T_n, T_n + L))$, and $\mathbf{C}$ is the covariance matrix of $F(T_i, T_i + L)$, $F(T_j, T_j + L)$'s.

6.2 Modeling Implied Volatility

Instead of learning option prices as discussed in Chap. 4, a strand of literature considers learning the implied volatility surface $\sigma_{Imp}(\cdot)$. Recall that given a (Call) option quote $P(T, \mathcal{K})$ with maturity T and strike $\mathcal{K}$, one may invert the Black-Scholes formula to find the scalar $\sigma_{Imp}(T, \mathcal{K})$ satisfying the equality $P(T, \mathcal{K}) = \mathrm{BSPrice}(T, \mathcal{K}, \sigma_{Imp})$. The respective volatility smile—the dependence of σ_{Imp} on option parameters—reflects deviations of the market from the Black–Scholes paradigm as the strike $\mathcal{K}$ and maturity T vary.

There are multiple contexts that motivate learning the full surface $\mathbf{x} \equiv (T, \mathcal{K}) \mapsto \sigma_{Imp}(T, \mathcal{K})$ given a training set of $(P_i)_{i=1}^N$. Note that one could proceed as in Chap. 4—building a surrogate $\hat{P}$ and then inverting to get $\hat{\sigma}_{Imp}(\mathbf{x}_*)$ from $\hat{P}(\mathbf{x}_*)$— or training directly on observed implied volatilities $\sigma_{Imp}^{1:N}$ [104]. The respective surrogate for σ_{Imp} enables transferring information obtained from vanilla Calls and Puts to a full-fledged asset model whereby one may price (with or without surrogates) arbitrary exotic options (i.e. options with non-standard payoffs, unlike Calls and Puts).

One motivating application is the Dupire formula that converts implied volatilities into a local volatility surface [67]:

$$\frac{\sigma_{loc}^2(T, \mathcal{K})}{2} = \frac{\partial_T \hat{P}(T, \mathcal{K}) + r\mathcal{K}\partial_{\mathcal{K}}\hat{P}(T, \mathcal{K})}{\mathcal{K}^2 \partial_{\mathcal{K}\mathcal{K}}^2 \hat{P}(T, \mathcal{K})}. \tag{6.11}$$

Observe that above the RHS is required to be positive, i.e. the surrogate $\hat{P}$ must be constrained to have nonnegative ∂_T temporal gradient and nonnegative $\partial_{\mathcal{K}\mathcal{K}}^2$ spatial convexity. To apply (6.11) one needs to know the full continuum of (inferred) option prices $\hat{P}(\cdot)$, leading to a calibration problem with the same flavor as term-structure construction. In the present context, probabilistic surrogates are appealing from a model risk perspective, yielding uncertainty quantification both on in-sample volatility points used for training, and for sampling of the entire local volatility surface forward in time.

A first attempt at using GPs for local volatility is done by Tegnér and Roberts [159] who placed a Gaussian prior directly on the local volatility surface σ_{loc}. To guarantee positivity, they assign a positive function on the prior. Such an approach leads to a nonlinear least squares training loss function, involving the nonlinear transformation of the local volatility into the corresponding vanilla option prices. Such a loss function is not amenable to gradient descent (stochastic or not), so the authors resort to a MCMC optimization. Tegnér and Roberts [159] use a GP prior with a SE kernel on the implied volatility and then a Markov Chain Monte Carlo algorithm to fit market prices. The prediction of the VIX index is also studied employing the same set-up. The choice of the SE kernel is driven by its smoothness: it is widely understood that a smooth volatility surface is desirable, especially for further hedging and pricing purposes. Moreover, the objective of minimizing reference option mispricing naturally connects to a likelihood model with additive observation noise. The noise variance (and more generally correlation structure) is then associated with the size of the bid-ask spread and it is naturally included in the Bayesian inference process.

For equity volatility surfaces, absence of arbitrage is guaranteed if there is no call spread, butterfly spread and calendar spread arbitrage opportunities. A GPR methodology, constrained on the no-arbitrage conditions for equity options, has been introduced in Chataigner et al. [29]. Using this approach, they achieved to produce arbitrage-free interpolated prices, and local volatility surfaces, with their uncertainties. This functional approach can be contrasted with a "stochastic" approach that seeks to find a close parametric, but arbitrage-free model, like SABR. The latter method requires several observed quotes at different strikes for each considered pair of tenors and maturities, which is actually not feasible in terms of liquidly quoted swaptions.

Qin and Almeida [138] fit a GP on the Black-Scholes implied log-volatilities,

$$\log \sigma_{Imp}(\mathbf{x}) \sim GP(0, k), \tag{6.12}$$

assuming constant observation noise. Hence the posterior implied volatility has a log-normal distribution. The market data consists of the mid-prices of SPX options from WRDS OptionMetrics database. Specifically, the authors use $\mathbf{x} = (M, \mathcal{K})$, with the forward moneyness $M \triangleq F/\mathcal{K}$, where $F = e^{rT} S_0$ is the forward price of the underlying at maturity T, and $\mathcal{K}$, the strike price specified in the option contract. The training set uses options with maturities of 20–365 days and moneyness in the range $M \in [0.7, 1.3]$, holding out non-round strikes as test set. The baseline is a quadratic model for σ_{Imp} from [49], the so-called ad-hoc or practitioner Black-Scholes (AHBS):

$$\sigma_{Imp}(M, T) = \left(a_0 + a_1 M + a_2 T + a_3 M^2 + a_4 T^2 + a_5 M \cdot T\right) \wedge 0.5 \vee 0.01.$$

The authors explore different model and kernel specifications to better exploit the underlying structure of the implied volatility function, aiming for a simple and intuitive method to nonparametrically calibrate σ_{Imp}, while also accounting for model uncertainty. Among kernels considered are the ARD versions operating on the scaled Euclidean distance $r^2(M, T) = (M - M')^2/(2\ell_M^2) + (T - T')/(2\ell_T^2)$ of Matérn-3/2 (2.10), M52, SE and Rational Quadratic (2.12) families, with the conclusion that the ARD k_{M32} provides the best results. Evaluation metrics are RMSE on the implied vol, maximum absolute deviation on the option price, and mean outside error which measures how much the estimated test option price is outside the bid-ask spread.

Stochastic Volatility

Learning the implied volatility mapping $\mathbf{x} \mapsto \sigma_{Imp}(\mathbf{x})$ fits naturally with GP regression. Less obvious, but intriguing connections also exist between GPs and *stochastic volatility* modeling which is concerned with describing the space-time dependence of the *instantaneous* volatility process $\sigma(t, \mathbf{x})$. Bennedsen et al. [16] uses the generalized Cauchy (2.14) kernel for this purpose due to its ability to simultaneously quantify both roughness and long-memory behavior. More broadly, the recent survey of Di Nunno et al. [47] includes rough volatility models, revealing several classes of GP priors appropriate for $\sigma(\cdot)$, many of which are Gaussian Volterra processes (recall Sect. 2.6).

6.3 Swaption Cubes

Cousin et al. [36] consider GPs for the swaption cube, which is a 3D object indexed by maturity, strike and tenor. This generalizes the implied volatility surface that is 2D, adding the third tenor dimension.

Let $S_w(T, T+t, \mathcal{K})$ be a receiver swaption with first payment at T, last payment at $T + t$ (hence a tenor of t) and strike $\mathcal{K}$. In the strike $\mathcal{K}$-dimension the butterfly and call-spread no-arbitrage conditions also apply to the swaption cube so that $\mathcal{K} \mapsto S_w(T, T + t, \cdot)$ must be a convex function. However the calendar spread does not hold. Instead, there is an additional arbitrage condition, named "in-plane triangular" inequality:

$$S_w(T, T + t, \mathcal{K}) + S(T + t, T + t + h, \mathcal{K}) \geq S(T, T + t + h, \mathcal{K}) \qquad (6.13)$$

for any tenors $t, t + h$. Compared to the equity price surface, the no-arbitrage constraints on swaption prices are non-directional. Indeed, the in-plane triangular inequality (6.13) involves prices of swaptions with different maturities and tenors. This constraint is more complex to handle than the calendar-spread no-arbitrage constraint in the equity option case. Moreover, constructing a constrained GP is much more involved in an input space of dimension three $\mathbf{x} \in \mathbb{R}^3$.

Cousin et al. [36] construct a finite dimensional approximation that is interpreted as a linear interpolation across a set of grid nodes $(T_i, t_j, \mathcal{K}_k)$ (cf. Sect. 4.4),

$$S(T, T + t, \mathcal{K}) = \sum_{i,j,k} S(T_i, T_i + t_j, \mathcal{K}_k)\phi_{i,j,k}(T, t, \mathcal{K}),$$

where ϕ is the respective hat function. A MVN structure with kernel k is then imposed on $\xi_{i,j,k} \equiv S(T_i, T_i + t_j, \mathcal{K}_k)$. For the latter GP on $(\xi_{i,j,k})$ they use a separable anisotropic kernel of the compound form

$$k = \eta^2 k_T \cdot k_t \cdot k_{\mathcal{K}}$$

with lengthscales $\ell_T, \ell_t, \ell_{\mathcal{K}}$. The separability allows to tensorize the GP covariance matrix and significantly reduces numerical complexity, moreover aiding MLE optimization. For the latter purpose, the authors explicitly compute the Jacobian of the log-likelihood function.

The results are illustrated with $N = 250$ training contracts with maturities $T = [5, 10, 15, 20]$ and tenors $t = [1, 2, 5, 10, 15, 20, 30]$. The authors then calibrate a SABR model to the GP-smoothed swap volatilities.

6.4 Mortality Rate Surfaces

Life actuaries are concerned with managing longevity risk, i.e. the risk of insureds passing away, which triggers cashflows (life insurance payouts, discontinuation of annuity payments, activation of survivor benefits, etc.). Longevity risk is also of great interest to demographers who study the evolution of a population subject to births, deaths and migration. In the standard framework, longevity is analyzed at population-level, meaning that individuals are aggregated into groups and the task is to consider overall, averaged behavior of the pool, described through *rates*. A mortality rate, also known as (instantaneous) force of mortality prescribes the rate at which a given "representative" individual in the pool might pass away. Using hazard rate paradigm, the survival probability from $t = 0$ until horizon T of an individual subject to mortality rate $f(t)$ is $\exp(-\int_0^T f(t)dt)$.

Mortality data is reported as pairs (D_t, E_t) representing the number of deceased and exposed within a given population pool and a given time window t, typically 1 year. Given D_t, E_t, the observed mortality rate is the ratio D_t/E_t and the typical models capture the log-mortality $y = \log D_t/E_t$.

Actuarial models focus on constructing dynamic life tables that reflect the dependence of mortality rates on age and calendar time. This leads to a gridded mortality surface indexed by Age x_{ag} and Year x_{yr}; fixing the underlying population and gender. As an illustration, the left panel of Fig. 6.2 displays the mortality surface for Danish Males, covering years 1990–2018 and Ages 50–84 grouped into 1-year buckets. Since mortality is roughly exponentially increasing in Age (known as the Gompertz Law), we plot the log-mortality. As expected, we observe an approximately linear (on the log-scale) trend in Age, as well as a decreasing trend across Years. One also records a lot of noise in the raw data: the precise number of deaths D_t each year arises from a confluence of a myriad of factors: individual health histories; environmental shocks like epidemics or heat waves; unexpected events like accidents or sudden deaths; etc. Therefore, the variance $\mathrm{var}(D_t|E_t)$ is significant.

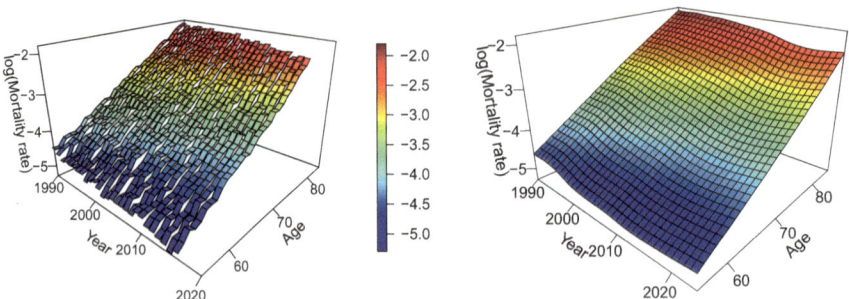

Fig. 6.2 Male mortality in Denmark. Left: raw age- and year-specific log mortality rate for the years 1990–2016 and ages 50–80. Right: GP-fitted mortality surface for years 1990–2024 and ages 55–85 trained on the former dataset and employing a separable M52 kernel

Stochastic mortality models aim to statistically de-noise available mortality data and then interpolate/extrapolate to make probabilistic predictions about mortality trends. This task is analogous to the previous constructions in this Chapter and fits well the GP paradigm. A GP model for mortality fits a surface $f_*(\cdot)$, indexed by the bivariate $\mathbf{x} \equiv (x_{ag}, x_{yr})$. One interprets f_* as the latent tendency of individuals in the pool to pass away, with realized mortality experience a noisy version of this "true" mortality. Predictions of $f_*(\mathbf{x}_*)$ correspond to de-noised forecasts, while predictions of $y(\mathbf{x}_*)$ can be converted into projections of future deceased counts.

A common choice for the kernel $k(\mathbf{x}, \mathbf{x}')$ is a separable product $k_{ag}(x_{ag}, x'_{ag}) \cdot k_{yr}(x_{yr}, x'_{yr})$. In contrast to other contexts, here the two coordinates of $\mathbf{x}$ have a different structure: mortality is generally smooth as a function of Age, but has rough, random-walk dynamics in Year. Therefore, different kernel families could be considered for k_{ag} and k_{yr}.

In [119], the proposal was for a product SE kernel, meaning

$$k(\mathbf{x}, \mathbf{x}') = k_{SE}(x_{ag}, x'_{ag}) \cdot k_{SE}(x_{yr}, x'_{yr}) =$$

$$\eta^2 \exp\left[-\frac{(x_{ag} - x'_{ag})^2}{2\ell_{ag}^2} - \frac{(x_{yr} - x'_{yr})^2}{2\ell_{yr}^2} \right], \qquad (6.14)$$

with the four hyperparameters: $\ell_{ag}, \ell_{yr}, \eta^2, \sigma_\epsilon^2$. The Age lengthscale ℓ_{ag} is interpreted as the strength of mortality dependence across ages. A typical $\ell_{ag} \in [10, 25]$ captures the idea of a demographic generation: two ages that are more than $2\ell_{ag}$ apart are effectively uncorrelated in their evolution. Demographers typically see a band of 15–30 years as a "generation" in terms of sharing similar life, health and demographic experiences. The Year lengthscale ℓ_{yr} represents strength of mortality correlation over years. Normally, $\ell_{yr} < \ell_{ag}$ reflecting more idiosyncratic fluctuations over time, understood as "more rough" temporal dynamics compared to the age structure.

Cohort Effect One special feature of mortality data is the so-called (birth) cohort effect, namely higher correlation among members of the same generation. To capture this extra dependence, it has been suggested to include a third term of the form $k_{coh}(x_{yr} - x_{ag}, x'_{yr} - x'_{ag})$ where $x_{coh} \equiv x_{yr} - x_{ag}$ in the kernel (6.14). For example, [86, 117] utilize separable Matérn-5/2 kernels, $k(\mathbf{x}, \mathbf{x}') = k_{M52}(x_{ag}, x'_{ag}) \cdot k_{M52}(x_{yr}, x'_{yr}) \cdot k_{M52}(x_{coh}, x'_{coh})$.

In a different vein, Debon et al. [43] consider a non-separable stationary Matérn space-time kernel originally due to Gneiting [71]. Let $u = |x_{yr} - x'_{yr}|$ be the distance in years and $h = |x_{ag} - x'_{ag}|$ and take

$$k(h, u) = \eta^2 \frac{1}{2^{\nu-1}\Gamma(\nu)(a|u|^{2\alpha} + 1)^{\delta+\beta/2}} \left(\frac{c\|h\|}{(a|u|^{2\alpha} + 1)^{\beta/2}} \right)^\nu$$

$$\times K_\nu \left(\frac{c\|h\|}{(a|u|^{2\alpha} + 1)^{\beta/2}} \right), \qquad (6.15)$$

where $K_\nu(\cdot)$ is the modified Bessel function of the second kind of order ν and the hyperparameters are $\boldsymbol{\theta} = (a, c, \alpha, \beta, \nu, \delta, \eta^2)$. The latter are fitted using a variogram and the median polish method.

Observation Likelihood In the main approach, the mortality GP is trained on log-mortality rates $y_i = \log[D(\mathbf{x}_i)/E(\mathbf{x}_i)]$ and the usual additive noise $\epsilon(\mathbf{x}_i)$ is then added as in (1.5). One may then utilize Gaussian likelihood, $\epsilon(\mathbf{x}_i) \sim \mathcal{N}(0, \sigma_\epsilon^2)$, e.g., treating σ_ϵ^2 as a hyperparameter. Alternatives exist, for example to impose a Poisson likelihood to capture the relationship between the observed number of deaths $D(\mathbf{x}_i)$ and the number of exposeds $E(\mathbf{x}^i)$:

$$D(\mathbf{x}) \sim Poi(e^{f(\mathbf{x})}E(\mathbf{x})). \tag{6.16}$$

As it turns out, mortality data is generally overdispersed relative to (6.16), so that its use (which requires additional approximations, cf. Sect. 3.2.3) is not necessarily better compared to the additive Gaussian noise [119]. The most promising approach is to utilize a fully Gaussian approach with input-dependent noise, setting $\sigma^2(\mathbf{x}) = \sigma_\epsilon^2 D(\mathbf{x})/E(\mathbf{x})$ [117]. This directly uses observable data and hence does not require either stochastic kriging or modeling of the $\mathbf{x} \to \sigma^2(\mathbf{x})$ surface.

Joint Models
Multi-output modeling is highly relevant for mortality analysis as one often needs joint models. This can be in order to model both genders (which experience high but not perfect correlation), or multiple populations; in the latter case one often augments the main population of interest with a reference population (e.g., one country with its neighbor(s), or a target insureds with the national experience). We refer to [86] for respective GP developments. Another extension is to model causes of death, which can be broken out at various degrees of granularity, see [87]. Additional extensions are joint models across geographical regions, such as U.S. states [116], as well as bespoke kernels for mortality [117].

6.4.1 Illustration: Danish Mortality

As an illustrative case study, we analyze male mortality in Denmark. The underlying data displayed in the left panel of Fig. 6.2 is Age-Year specific "1 × 1" from the publicly available Human Mortality Database [127], representing the official national-level mortality experience. We construct a GP model with a separable stationary kernel that is a product of univariate Matérn-5/2 kernels in Age and in Year, and a linear prior trend in Age, $\mu(\mathbf{x}) = \beta_0 + \beta_{ag}^1 x_{ag}$. The GP hyperparameters are optimized using MLE on the training set of $ag \in \{60, \ldots, 80\}$ and $yr \in$

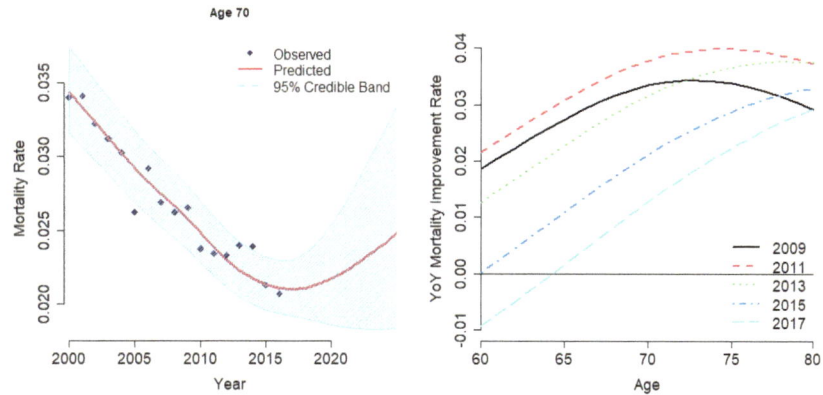

Fig. 6.3 Left: GP forecasts of Male mortality in Denmark at Age 70 across calendar years. Right: annual mortality improvement factors as a function of Age

{1990, ..., 2016} which contains a total of $21 \cdot 27 = 567$ inputs. The fitted hyperparameters are $\beta_0 = -9.852$, $\beta_{ag}^1 = 0.0919$, $\ell_{ag} = 24.28$, $\ell_{yr} = 9.88$, $\eta^2 = 0.0369$ and $\sigma_\epsilon^2 = 0.00179$. The coefficient β_{ag}^1 means that mortality increases on average by 9.19% for each additional year of Age.

The left panel of Fig. 6.3 shows a one-dimensional view of the raw and fitted Danish mortality across time, fixing $x_{ag} = 70$ and varying x_{yr}. In this dataset the observations stop at 2016, so years 2017–2023 are out-of-sample projections. We observe a clear downward trend, i.e. the GP model predicts a continuation of mortality improvement, although the uncertainty cone–obtained as a 95% predictive band around the centered forecast is quite large. Note that what is displayed is the GP-based uncertainty (as a 95%-level interval of width $\pm 1.96 s(\mathbf{x}_*)$ around $m(\mathbf{x}_*)$, cf. Chap. 1) of the latent mortality rate in the future; there is additional separate noise capturing the aforementioned "intrinsic" fluctuations in the realized death counts. The plot also shows the reversion to the prior as x_{yr} increases and the impact of conditioning on data shrinks, illustrated by the widening prediction intervals and the upward pull of $m(\mathbf{x}_*)$ beyond year 2020. The speed of this reversion is driven by the lengthscale parameter $\ell_{yr} - 9.88$ which captures how far out into the future does the recent mortality experience still carry influence.

Mortality evolves very slowly over time. Therefore, rather than forecasting mortality per se, life actuaries often focus on the *mortality improvement factors*, which is the rate of change of mortality over time, $\partial_{yr} f_*(\cdot)$ (in practice implemented as an annual finite difference). A typical MI of 0.02 means that mortality decreases by 2% (in relative, not absolute terms) per year. The right panel of Fig. 6.3 shows mortality improvement in Danish Males in the 2010s, demonstrating that mortality gains have slowed down over the decade. We observe that the MI is highly age-dependent: the trend at age 60 is of worsening mortality (MI below zero) while at age 80 it is improving at more than 2.5% per year.

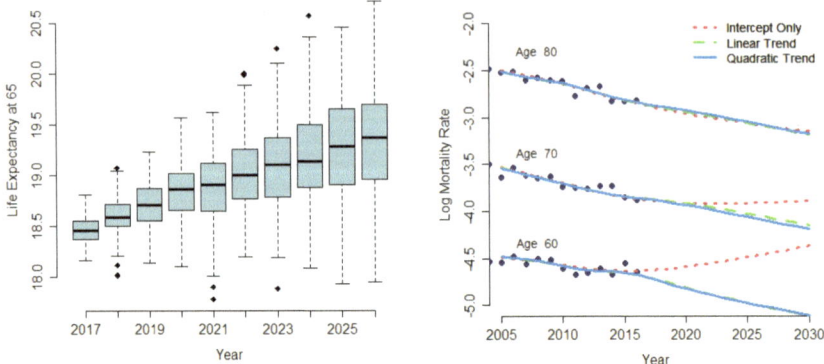

Fig. 6.4 Left: predictive distribution of future period life expectancy e_{65} for Danish males aged 65. Right: Impact of mean function $\mu(\cdot)$ choice on Denmark male log-mortality forecasts over the period 2005–2030

As a final application, the GP model can be used to construct probabilistic forecasts of actuarial metrics, such as life expectancy. Period life expectancy (LE) at Age ag is the mean length of life of a hypothetical cohort assumed to be exposed to the mortality rates observed at a given year,

$$e_{ag,yr} \triangleq \sum_{s>ag} (s - ag) \cdot \mathbb{P}(\tau = s) = \sum_{s} e^{f_*(s,yr)} \cdot (s - ag) \cdot \exp\left(-\sum_{u=t}^{s} e^{f_*(u,yr)}\right)$$

where τ is the remaining lifetime. GPs allow to capture the dependence structure of future rates $f_*(\cdot, yr)$ which are highly correlated. The left panel of Fig. 6.4 shows boxplots illustrating the LE $e_{65,yr}$ for future $yr \in \{2017, 2026\}$, obtained by drawing 200 samples of the GP-fitted $f_*(\cdot, yr)$. Reflecting the nature of GP posterior bands, the Figure shows a rapidly growing cone of uncertainty regarding the trend in $e_{65,\cdot}$ as a function of time.

Mean Function Modeling   As mentioned, mortality rates exhibit unmistakeable trends in Age and possibly in Year. This feature of mortality data suggests the incorporation of a non-constant prior mean $\mu(\mathbf{x})$, allowing the GP to focus on modeling less-understood structure. The right panel of Fig. 6.4 demonstrates GP fits (based on universal kriging formulas (1.27) and (1.28)) using three different prior means: (i) $\mu(\mathbf{x}) = \beta_0$ constant; (ii) $\mu(\mathbf{x}) = \beta_0 + \beta_{yr}^1 x_{yr} + \beta_{ag}^1 x_{ag}$ linear in both Age and in Year; (iii) $\mu(\mathbf{x}) = \beta_0 + \beta_{yr}^1 x_{yr} + \beta_{ag}^1 x_{ag} + \beta_{ag}^2 x_{ag}^2$ which is quadratic in Age an linear in Year. In the figure we show the resulting predictions up to 2030. The constant prior mean fit includes a strong upward trend especially for Age 60 which is the direct consequence of mean-reversion: as $x_{yr} \to \infty$, we have that $m(\mathbf{x}) \to \mu(\mathbf{x}) = \beta_0 = -3.728$. As a result the GP predicts that in the distant future, all Ages will experience the baseline mortality of $\exp(-3.728) = 2.4\%$ per year. As Fig. 6.4 shows, this "bug" is actually not just about the distant future, but already greatly influences the GP prediction just 10–12 years beyond training data. The conclusion is that $\mu(\mathbf{x}) = \beta_0$ is *not* an appropriate pick if our goal is to indeed extrapolate and forecast future mortality. But note that a constant prior mean is quite benign for in-sample smoothing: for $x_{yr} < 2015$ one can barely distinguish among the three curves.

Remark 6.2 Like in the case study of Sect. 1.4.3 , de-trending lowers the length-scales of the GP modeling the residual, and the respective outputscale η^2. For example, for the constant prior mean we find that $\hat{\eta}^2_{(i)} = 1.4$ while for the quadratic prior mean we get $\hat{\eta}^2_{(iii)} = 0.0011$, more than a 1000 times smaller.

The BICs (2.25) of these three GP models are 616.28, 627.65, 626.27, which indicates that the constant prior mean is clearly deficient (Bayes Factor (2.26) of about 10^{-5} compared to model (iii)), while there is minimal preference for the linear prior compared to a quadratic one (BF of 3.97). Observe that model (i) has a single trend coefficient β_0, while (ii) has two more hyperparameters and (iii) has three more $(\beta^1_{yr}, \beta^1_{ag}, \beta^2_{ag})$, with those extra degrees of freedom penalized accordingly in the above BICs.

6.5   Valuation of Variable Annuities

Variable annuities (VAs) is a class of long-term financial contracts that serve dual roles: investment vehicles and life insurance products. VAs are marketed as enhancements of life insurance, offering the upside of investment opportunities, making them "variable," while also providing insurance guarantees in the form of (minimum) payments linked to the insured's life status. The VA policyholder pays an initial premium to the insurer, in exchange for several future benefits and features, such as:

- Death Benefit—Minimum payment to the beneficiary upon death;
- Account Benefit—Minimum account value after a specified period;
- Income Benefit—Minimum annuity income after a set period;
- Maturity Benefit—Minimum account value at contract maturity;
- Withdrawal Benefit—Annual withdrawals up to a limit;
- Mixed Benefits—Combination of different benefit types;
- Return of Premium—Minimum benefit based on the premium;
- Annual Roll-Up—Benefit increases at a predetermined rate;
- Annual Ratchet—Benefit resets if current account value exceeds it.

These options can be combined; for instance, in a DBRU plan, the death benefit (DB) is tied to market performance, and the benefit rate increases annually (RU).

Calculating the fair market valuation of a VA contract is computationally demanding, since the relevant financial inputs encapsulate investment risk factors, contract type and embedded guarantees and policyholder mortality factors like age and gender. An insurance portfolio consists of thousands of VAs that vary in the amounts and types of the embedded guarantees.

For example, the synthetic dataset in [62] consists of $N = 190,000$ instances of different VA contracts indexed by 26 features $\mathbf{x} \in \mathbb{R}^{26}$. Notably, some of these features are *categorical*, encoded using one-hot encoding (with baseline). For instance the policyholder gender can be encoded as $x_{gndr} = 0$ for Females and $x_{gndr} = 1$ for Males.

The typical procedure to value a single VA contract is via Monte Carlo simulations, which becomes prohibitive for such large N. As a result, surrogate models that provide a statistical prediction for VA valuation is an attractive solution. Representing a single portfolio as $(\mathbf{x}, y)$ with y signifying the estimated VA market value with features $\mathbf{x}$, one may construct a GP f for this purpose.

Given a database $\mathcal{D}'$ of N' VAs, the training stage of fitting f_* is known as *scenario reduction*, aiming to find N_{tr} representative contracts that maximize the accuracy of the GPR predictions on the full N' set. One approach is to use LHS to provide a space-filling subsampling, i.e. to select $\mathcal{D}$ that is equidistributed across the input space $\mathcal{X}$. To this end, one first generates a space-filling $\mathcal{D}_{LHS}$, of N_{tr} locations via regular LHS. Each tuple $\mathbf{x} \in \mathcal{D}_{LHS}$ is then mapped to its nearest-neighbor in the full dataset $\mathcal{D}'$ based on Euclidean distance, yielding a representative collection $\mathcal{D}$ of contract inputs.

Given a high-dimesional $\mathcal{X}$, the recommendation for the GP covariance, is to use a separable SE product kernel with Automatic Relevance Determination (ARD):

$$k(\mathbf{x}, \mathbf{x}') = \prod_{j=1}^{d} k_j(x_j, x'_j), \qquad (6.17)$$

where k_j denotes the base kernel associated with the j-th dimension. The idea of ARD is to assign a unique lengthscale parameter, ℓ_j, to each dimension. During MLE optimization, feature relevance is assessed by the magnitude of ℓ_j. Features that are deemed irrelevant by the GP model end up with a large ℓ_j which renders $k_j(x_j, x'_j)$ close to one, consequently diminishing its influence in the kernel computation (6.17). For example, with the SE kernel the lengthscale ℓ_j scales the exponential term:

$$\lim_{\ell_j \to \infty} k_{SE}(x_j, x'_j) = \lim_{\ell_j \to \infty} \exp\left(-\frac{(x_j - x'_j)^2}{2\ell_j^2}\right) = 1,$$

illustrating the nullification of dimension j's influence when its lengthscale ℓ_j approaches infinity. For the binary categorical features, the one-hot encoding acts as a scaling factor in the covariance: if the category of $\mathbf{x}, \mathbf{x}'$ matches, the corresponding term in the product (6.17) becomes $k_j(1, 1) = k_j(0, 0) = 1$ and if not, it becomes $k_j(0, 1) = k_j(1, 0) = \exp(-0.5/\ell_j^2) < 1$, scaling the contribution of that feature.

ARD offers a convenient way to quantify variable importance, essential to interpret which features drive the VA values especially when many of them are collinear.

Goudenège et al. [76] focus on GMWB VA contracts in the context of a Heston-Hull-White model for stochastic volatility and interest rates respectively, emphasizing the sensitivity to market model (rather than VA) parameters. In their setting, the primary factors is the account value (A_t), the respective volatility (v_t) and the interest rate (r_t), satisfying

$$\begin{cases} dA_t = (r_t - \alpha)A_t dt + \sqrt{v_t} A_t dW_t^S; \\ dv_t = \kappa_v(\theta_v - v_t) + \eta_v \sqrt{v_t} dW_t^v; \\ dr_t = \kappa_r(\theta_r(t) - r_t)dt + \eta_r dW_t^r \end{cases}$$

where α are the VA fees, and with correlations $d\langle W^r, W^S \rangle = \rho_r dt$ and zero correlation $d\langle W^v, W^r \rangle$. The resulting GP input $\mathbf{x} \in \mathbb{R}^{11}$ consists of the 9 model parameters $v_0, \kappa_v, \theta_v, \eta_v, \rho_v, r_0, \kappa_r, \eta_r, \rho_r$ and the two VA parameters α, κ (fees and early withdrawal penalty). Of note, $\mathbf{x}$ does not include maturity T, instead a separate GP is built for each T. Goudenège et al. [76] train an anisotropic SE GP, based on a space-filling Faure simulation design, uniform in each coordinate. Training data $\mathbf{y}$ is generated from a hybrid tree-PDE 1D solver. As applications, [76] use the GP surrogate f_* to solve for the fair fee α that makes the contract value to equal to its given premium P, as well as to estimate the Delta $\partial_A f_*(\mathbf{x})$ with respect to the account value A_t of the contract (the Delta is estimated via a separate GP, not taking gradients).

Further Reading

The non-mortality applications in this Chapter are primarily based on [29, 36]; the mortality Section is based on [117, 119]. The monograph by Gan and Valdez [63] provides more detail about VA surrogates, including description of the synthetic dataset and accompanying R code for implementation.

This Chapter is accompanied by (i) a R Markdown notebook that illustrates the basics of fitting a mortality surface using a GP surrogate with a separable kernel, applied to Danish Males dataset and (ii) a Python Jupyter notebook illustrating fitting a GP with an additive kernel to a single end-of-day forward curve of natural gas futures quotes, see Sect. 6.1.1.

Chapter 7
Stochastic Control

GPs can be used as emulators for more general stochastic control problems. In a nutshell, the idea is to extend the emulation of the q-value in Chap. 5 to other settings. In this Chapter we outline some of the extant applications of GPs in switching control, continuous control and impulse control. Our exposition covers different flavors of Gaussian Process Dynamic Programming (GPDP) that is based on value function approximation techniques for stochastic control.

7.1 Switching Control

Switching control problems generalize optimal stopping to allow for multiple system regimes that the controller may repeatedly choose among. The key feature of optimal switching is the regime $j \in \mathcal{J}$, that affects the dynamics of the state process $(\mathbf{X}_t)$ and/or the corresponding running payoffs (g_t). A motivating example is on/off control, where an agent aims to optimally toggle in and out of the active state depending on the state $\mathbf{X}_t$, for instance to start/stop a power plant depending on electricity price(s) $(\mathbf{X}_t)$ or to toggle the operations of a mine depending on metal price $(\mathbf{X}_t)$.

Discretizing the state dynamics, an optimal switching problem considers a vector-valued $(\mathbf{X}_k)$ evolving according to the transition map

$$\mathbf{X}_{k+1} = \Psi(t_k, \mathbf{X}_k, j_k, \epsilon_{k+1}),$$

© The Author(s), under exclusive license to Springer Nature Switzerland AG 2025   115
M. Ludkovski, J. Risk, *Gaussian Process Models for Quantitative Finance*,
SpringerBriefs in Quantitative Finance,
https://doi.org/10.1007/978-3-031-80874-6_7

where ϵ_k are exogenous independent shocks in time step k, with the objective of maximizing the finite-horizon cumulative reward

$$\mathbb{E}\left[\sum_{k=0}^{K-1}[g(\mathbf{X}_k, j_k) - \kappa(j_k, j_{k+1})] + G(\mathbf{X}_T, j_T)\right] \tag{7.1}$$

where $\kappa(j, j') \geq 0$ are the switching costs that penalize actions, g is the step-wise profit at step t_k and G is the terminal payoff at $T = t_K$. The resulting value function $V(t, \mathbf{x}, j)$ for (7.1) now depends on the initial switching regime j and satisfies the DPP

$$V(t_k, \mathbf{x}, j) = \max_{\ell \in \mathcal{J}} \mathbb{E}\left[g(\mathbf{x}, j) - \kappa(j, \ell) + V(t_{k+1}, \mathbf{X}_{k+1}, \ell)|\mathbf{X}_k = \mathbf{x}\right], \tag{7.2}$$

where the expectation is over the external shock ϵ_{t_k}. Hence optimal switching is reduced to the local optimization over potential new regimes $\ell \in \mathcal{J}$, taking into account the look-ahead $V(t_{k+1}, \cdot)$. Leveraging again the Markov property, observe that the optimal action $j_k^*(\mathbf{x}, j)$ can be thought of as a discrete-valued feedback map, generalizing the concept of the stopping set $\mathcal{S}$ in Chap. 5 to the task of partitioning the state space $\mathcal{X}$ into the disjoint switching regions $\mathcal{S}_k(j, \ell) \stackrel{\triangle}{=} \{\mathbf{x} \in \mathcal{X} : j_k^*(\mathbf{x}, j) = \ell\}$. As an example, [115] considered valuation of natural gas storage facilities where $|\mathcal{J}| = 3$ corresponds to injection, withdrawal and do-nothing regimes, and the state is bivariate $\mathbf{X} = (P, I)$ representing the price of gas and the storage inventory. For storage problems, the external shock is univariate $\epsilon_{k+1} \in \mathbb{R}$ and only affects the new P_{k+1}, while $I_{k+1} = I_k + c(j_k)\Delta t$ has deterministic dynamics.

As one approach to a numerical scheme for finding $V(t_k, \cdot, \cdot)$, one may generalize the RMC framework by separately emulating $\mathbf{x} \mapsto q(t_k, \mathbf{x}, j) \stackrel{\triangle}{=} \mathbb{E}[V(t_{k+1}, \mathbf{X}_{k+1}, j)|\mathbf{X}_k = \mathbf{x}]$ for each regime j and then directly setting

$$j^*(\mathbf{x}, j) = \arg\max_{\ell \in \mathcal{J}} \{\widehat{q}(t_k, \mathbf{x}, \ell) - \kappa(j, \ell)\}.$$

This effectively follows the same recipe as in Chap. 5 to learn (5.7). Ludkovski and Maheshwari [115] discuss implementation of such a scheme through GPs, applying an SE kernel in $d = 2$. They also address different training designs, including space-filling strategies and adaptive designs tailored to the storage valuation problem. Hu and Ludkovski [85] address more sophisticated sequential design heuristics that aim to sequentially learn the maximizer of $(\mathbf{x}, j) \mapsto q(t_k, \mathbf{x}, j)$, i.e., emphasizing learning the switching regions rather than the q-values themselves, similar to the adaptive methods in Sect. 5.2.

7.2 Continuous Control

Let $\mathcal{T} = \{t_k : k = 0, 1, 2, \ldots, K, \ t_K = T\}$ be the discrete time set with expiration time T. The stochastic continuous control problem replaces $j_k \in \mathcal{J}$ with a continuously-valued control $u_k \in U$. Globally, the optimization now takes places over the family $\mathcal{U}$ of $\mathbb{F}$–adapted admissible feedback control processes $u \overset{\triangle}{=} (u_t, \ t \in \mathcal{T})$. In the continuous control setting, the dynamics remain the same, replacing j with u

$$\mathbf{X}_{k+1} = \Psi(t_k, \mathbf{X}_k, u_k, \epsilon_{k+1}), \tag{7.3}$$

where Ψ captures the joint (time-dependent) effect of the current state $\mathbf{X}_k$, the action u_k and the exogenous shock ϵ_{k+1} that is assumed independent across time steps.

The objective is again maximizing total finite-horizon reward on $[0, T]$, now with respect to $u \in \mathcal{U}$

$$V(t_0, \mathbf{x}) \overset{\triangle}{=} \sup_{u \in \mathcal{U}} \mathbb{E}\left[\sum_{k=0}^{K-1} g(t_k, \mathbf{X}_k, u_k) + G(\mathbf{X}_T) \,\Big|\, \mathbf{X}_0 = \mathbf{x}\right]. \tag{7.4}$$

This dynamic optimization problem is reduced to the DPP-based Bellman equation

$$V(t_k, \mathbf{x}) = \sup_{u \in U} \left\{g(t_k, \mathbf{x}, u) + \mathbb{E}\left[V(t_{k+1}, \Psi(t_k, \mathbf{x}, u, \epsilon_{k+1})) \,\big|\, \mathbf{X}_k = \mathbf{x}\right]\right\} \tag{7.5}$$

with terminal condition $V(T, \mathbf{x}) = G(\mathbf{x})$. The backwards recursive solution is similar to that in Sect. 5.1, where for $k = K - 1, \ldots, 1, 0$, and assuming the supremum is attained, one considers the state-action q-functions

$$q(t_k, \mathbf{x}, u) = g(t_k, \mathbf{x}, u) + \mathbb{E}\left[q(t_{k+1}, \Psi(t_k, \mathbf{x}, u, \epsilon_{t_{k+1}}), u) \,\big|\, \mathbf{X}_k = \mathbf{x}\right]$$
$$u^*(t_k, \mathbf{x}) = \arg \sup_{u \in U} q(t_k, \mathbf{x}, u), \tag{7.6}$$

and consequently $V(t_k, \mathbf{x}) - q(t_k, \mathbf{x}, u^*(t_k, \mathbf{x}))$. Gaussian Process Dynamic Programming (GPDP) [31] is based on a value function approximation $\widehat{V}$ for V, providing an entry point for employing a GP surrogate.

Solving (7.5) requires not only learning the functional approximation for the multi-dimensional map $\mathbf{x} \mapsto V(t_k, \mathbf{x})$ or $(\mathbf{x}, u) \mapsto q(t_k, \mathbf{x}, u)$ which necessarily can only be sampled, but two additional steps. First, one needs to handle the integration over ϵ_{k+1} in (7.5), with the integrand being the (recursively constructed) $V(t_{k+1}, \cdot)$. Second, one must carry out the numerical optimization for $u^*(t_k, \mathbf{x})$ which is generally not available analytically and is given implicitly in terms of first-order-conditions involving $q(t_k, \mathbf{x}, \cdot)$. The latter step suggests building a second surrogate $\widehat{u}^*(t_k, \mathbf{x})$.

In Chen and Ludkovski [31], the strategy is to generate two sequences of GP surrogates, based on pointwise training outputs v_k^i, $\breve{u}_k^i$ at training inputs $\mathbf{x}^i$, $i = 1, \ldots, N$. The conditional expectation over ϵ_{k+1} in (7.5) is approximated with a weighted average $\hat{E}[\cdot | \mathbf{X}_k]$ based on quantizing (i.e., replacing with a finite sum) the distribution of ϵ_{k+1}, so that given $\widehat{V}(t_{k+1}, \cdot)$,

$$v_k^i \stackrel{\triangle}{=} \max_{u \in U} \left\{ g(t_k, \mathbf{x}^i, u) + \hat{E}\left[\widehat{V}\left(t_{k+1}, \Psi(t_k, \mathbf{x}^i, u, \cdot) \right) \right] \right\} \tag{7.7}$$

and $\breve{u}_k^i$ is the corresponding arg sup of (7.7). These populate $\mathcal{D}_k^V = \{ (\mathbf{x}^i, v_k^i) : i = 1, \ldots, N \}$ and $\mathcal{D}_k^u = \{ (\mathbf{x}^i, \breve{u}_k^i) : i = 1, \ldots, N \}$ which consequently produce GP surrogate fits for the value function $\widehat{V}(t_k, \cdot)$ and the optimal control $\widehat{u}(t_k, \cdot)$ at decision time t_k. As usual for making predictions, the posterior mean of $\widehat{V}(t_{k+1}, \cdot)$ is utilized.

Note that this approach first fixes $\mathbf{x} \in \mathcal{X}$ and then performs the local maximization over U to obtain $\breve{u}_k(\mathbf{x})$ and eventually the surrogate $\mathbf{x} \mapsto \widehat{u}(t_k, \mathbf{x})$. Alternatively, Kharroubi et al. [95] reverse the order of optimization and emulation, first evaluating the state-action q-function by building the surrogate $(\mathbf{x}, u) \mapsto \widehat{q}(t_k, \mathbf{x}, u)$ and consequently maximizing it to recover $\widehat{u}(t_k, \mathbf{x})$. Besides being more direct, (7.7) maintains a lower input dimension for $\widehat{V}(t_k, \cdot)$ compared to that of $\widehat{q}(t_k, \cdot, \cdot)$.

The pointwise optimization in (7.7) involves a function of the GP posterior mean $\widehat{V}(t_{k+1}, \cdot)$. Since the latter is data-driven, the problem is generally non-convex and can be nontrivial to solve even in the setting when the action space U is one-dimensional. A default choice is to use a gradient-free optimizer, which however can be prohibitively slow for multi-dimensional control spaces. The hinted preference for gradient-based solvers (such as the commonly applied L-BFGS scheme) can be accommodated by using the analytic expressions for GP gradients, directly passing in $\nabla \widehat{V}(t_{k+1}, \cdot)$. To this end, the recommendation is to use a smooth-enough GP for $\widehat{V}$, such as SE (2.6) or Matérn-5/2 (2.11) kernels, rather than non-smooth kernels like k_{M32} or k_{M12}. On the other hand, the control GP $\widehat{u}$ tends to be more rough than the value function, and hence might be better captured with a M32 or M12 kernel. See [10] and [31] for further details. A further technique to regularize the optimization in (7.7) and avoid getting trapped in local maxima is to take into account additional information about $\widehat{V}$, for example its posterior variance.

The original version of GPDP appeared in Deisenroth et al. [44] in the context of reinforcement learning. The authors consider a general discrete-time stochastic control problem; starting with the training set $\mathcal{D}_k$, which they call the *support points*, they propose to first train independent GPs for the pointwise q-value

$$u \mapsto \mathbb{E}[g(t_k, \mathbf{x}^i, u) + \widehat{V}(t_{k+1}, \Psi(t_k, \mathbf{x}^i, u, \epsilon_{k+1})) | \mathbf{X}_k = \mathbf{x}^i] \simeq \widehat{q}^{(i)}(u)$$

for each $\mathbf{x}^i \in \mathcal{D}_k$. After numerically optimizing $\widehat{q}^{(i)}(\cdot)$ to obtain $\breve{u}_k^i$ and computing the corresponding v_k^i they then train the global GP based on $((\mathbf{x}^i, v_k^i))_{i=1}^N$ for $\widehat{V}(t_k, \cdot)$. This scheme avoids building a global state-action surrogate, instead employing a family of N GPs indexed by $\mathbf{x}^i$. The motivation is that q might be discontinuous in $\mathcal{X}$. Moreover, this strategy facilitates creating an adaptive design $u^{1:N_u}$ for learning $\widehat{q}^{(i)}(\cdot)$ with the idea being to train around the maximizer $u^*(t_k, \mathbf{x}^i)$ rather than across the entire action space U. This framework also crystallizes the idea that the $\widehat{q}^{(\cdot)}$ surrogates exist for the purposes of interpolating within the continuous action space U while the $\widehat{V}$ surrogate interpolates in the continuous input space $\mathcal{X}$. As a third step, [44] train GPs for the policy map $\widehat{u}$. In some of their case studies, the ground truth $u^*(\cdot)$ is discontinuous (e.g. either positive or negative, but always bounded away from zero), with the approach being to train multiple GPs (essentially a partitioned GP like in Sect. 3.4) for each piece and then add a classifier-GP to select the appropriate case.

Training Across Time-Steps
Another theme in using GPs for Dynamic Programming is that one must fit a whole sequence of GP models indexed by the time step k. From a computational perspective, this implies that a significant portion of the algorithm running time is spent on inferring the respective hyperparameters $\boldsymbol{\theta}_k$. Generally, $\boldsymbol{\theta}_k$ is stable across time, offering the opportunity to speed up the MLE optimization by warm-starting it with the previous-step values, or directly taking $\boldsymbol{\theta}_k = \boldsymbol{\theta}_{k+1}$, freezing hyperparameters across (some) time-steps. Empirically this is often an acceptable speed-up: because the GP surrogate is data-driven, the resulting prediction depends mostly on the actual training data $\mathcal{D}_k$ and less on the precise value of the hyperparameters $\boldsymbol{\theta}_k$.

7.3 Impulse Control

Stochastic impulse control is concerned with systems where the state process $(\mathbf{X}_t)$ is subject to stochastic dynamics, as well as repeated interventions by the controller that make an instantaneous impact on $(\mathbf{X}_t)$ and carry an instantaneous cost/reward. The goal of the controller is to maximize total expected (discounted) profit that is driven by the impulses z_m's and the running revenue function $g(t, \mathbf{X}_t)$. Impulse controls are common to describe management of natural resources and industrial capacity planning, such as capacity expansion [17, 56], stochastic harvesting problems [13], control of foreign exchange rates [24] and management of energy retail prices [11].

Due to having many ingredients to solve for: impulse thresholds, impulse targets, intervention function, value function, etc., only very special cases, such as time-stationary models with linear intervention costs and linear dynamics can be solved explicitly, otherwise necessitating numerical methods.

Impulse control can be viewed as a generalization of multiple optimal stopping. Namely, at each time instant, the controller must first decide whether to act or to wait; conditional on acting, in the second decision stage the controller picks the optimal action. This perspective reduces impulse control to repeated optimal stopping with an implicit payoff function specified via the so-called intervention operator $\mathcal{M}$. Through this lens, solvers for impulse control can be built on top of related code for optimal stopping after incorporating the computation of the intervention operator.

Below for simplicity we describe a one-dimensional state process (X_t) in discrete time with decision grid $\mathcal{T} = ((t_k)_{k=0}^K)$ with equal spacing Δt. Without interventions, X follows the discrete dynamics

$$X_{k+1} = \Psi(X_k, \epsilon_{k+1}),$$

for example as the discretization of a stochastic differential equation $dX_t = a(X_t)\,dt + b(X_t)\,dW_t$. If an intervention takes place at step k, then X is instantaneously moved to $X_k + z$ (a displacement by z) and simultaneously an intervention cost $\kappa(t_k, X_k, z)$ is assessed. The space of interventions is denoted $z \in \Xi$.

An admissible impulse strategy $\mathbf{z} \in \mathfrak{Z}$ is recorded as the sequence of intervention times τ_m (which are $\mathbb{F}$-stopping times) and respective impulses z_m, so that $\mathbf{z} = (\tau_1, z_1, \dots,)$. Given a horizon T and terminal condition $G(X_T)$, the goal is to evaluate the *value function* $V : [0, T] \times X \to \mathbb{R}$, quantifying optimal expected reward

$$V(t, x) \overset{\triangle}{=} \sup_{\mathbf{z} \in \mathfrak{Z}_t} \mathbb{E}\left[\sum_{k=0}^K g(k, X_k) - \sum_{m:\tau_m < T} \kappa(\tau_m, X_{\tau_m}, z_m) + G(X_T) \,\Big|\, X_k = x \right],$$
$$(7.8)$$

where $g : (t, x) \mapsto \mathbb{R}$ is the running reward function, and $\kappa : (t, x, z) \mapsto \mathbb{R}$ is the cost of applying impulse of size $z \in \mathbb{R}$ at time t and state x. This resembles continuous control except for the special way of assessing the intervention costs and the case-wise dynamics of X: $X_{k+1} = \Psi(X_k, \epsilon_{k+1})1_{\{k \notin \{\tau_1, \dots\}\}} + \sum_m \Psi(X_k + z_m, \epsilon_{k+1})1_{\{k=\tau_m\}}$.

The dynamic programming Bellman equation for impulse control on $[t_k, t_{k+1}]$ is:

$$V(t_k, x) = g(t_k, x) + \max(q(t_k, x), \mathcal{M}(t_k, x)),\qquad(7.9)$$

where $q(t_k, x) \overset{\triangle}{=} \mathbb{E}\left[V\left(t_{k+1}, \Psi(X_k, \epsilon_{k+1})\right) \mid X_k = x \right]$ is the familiar look-ahead q-value and

$$\mathcal{M}(t_k, x) \overset{\triangle}{=} \sup_{z \in \Xi}\{q(t_k, x + z) - \kappa(t_k, x, z)\}.$$

The latter *intervention operator* captures the value of making the best possible impulse at t_k. In sum, at each time period k, the controller must decide whether to continue (0) or act ($\neq 0$). In the latter case, she needs to further find the best action z^*. This matches the appearance of the max in (7.9)—one should continue if the q-value dominates the intervention value, and one should impulse otherwise. Within a Markovian structure an impulse strategy $\mathbf{z}$ can be equivalently encoded as the collection of feedback *action maps* $\mathcal{Z}_k(x) \in \{0\} \cup \Xi$:

$$\mathcal{Z}_k(x) = \arg\sup_z\{q(t_k, x + z) - \kappa(t_k, x, z)\} \cdot 1_{\{\mathcal{M}(t_k, x) > q(t_k, x)\}}. \tag{7.10}$$

The action map $\mathcal{Z}_k$ gives rise to the action region $\mathfrak{S}_k \triangleq \{x : \mathcal{Z}_k(x) \neq 0\} \subseteq \mathcal{X}$, where the optimal choice is to act.

Regression Monte Carlo for impulse control proceeds by recursively constructing surrogates

$$\widehat{q}(t_k, \cdot) \simeq \mathbb{E}\left[\widehat{V}(t_{k+1}, X_{k+1}) \middle| X_k = \cdot\right]$$

that proxy for $q(t_k, \cdot)$ and are used to induce the respective $\widehat{\mathcal{Z}}_k$ according to (7.10). The resulting scheme, described in Algorithm 1 reduces optimal impulse control to a double sequence of probabilistic function approximation tasks. The primary task entails fitting a functional approximator $\widehat{q}_k$ based on empirical pointwise values $y_{k+1}^{1:N_k}$. The secondary task is to learn the intervention operator

$$\widehat{\mathcal{M}}(t_k, \cdot) = \sup_{z \in \Xi}\{\widehat{q}(t_k, \cdot + z) - \kappa(t_k, \cdot, z)\} = \widehat{q}(t_k, \cdot + \widehat{\mathcal{Z}}_k(\cdot)) - \kappa(t_k, \cdot, \widehat{\mathcal{Z}}_k(\cdot)) \tag{7.11}$$

approximating $\mathcal{M}(t_k, x)$. The proposal is then to employ GP models, training the GP for $\widehat{q}(t_k, \cdot)$ within a backward recursion across time steps $k = K - 1, \ldots, 1$, in full analogue to the continuous control and optimal stopping settings. Of note, Algorithm 1 generates the training outputs $y^{1:N_k}$ through a w-step forward simulation, first collecting realized rewards and intervention costs over the time steps $\{k + 1, \ldots, k + w\}$ and then using $\widehat{q}(k + w, \cdot)$ to estimate the remaining future reward; see [45, 113] for further details.

The ultimate expected reward can be obtained as the sample average reward across a fresh set of $\{x_k^{n', \widehat{\mathcal{Z}}}, k = 1, \ldots, K, n' = 1, \ldots, N', x_0^{n'} = x\}$ controlled trajectories,

$$\check{V}(0, x) = \frac{1}{N'} \sum_{n'=1}^{N'}\left\{\sum_{k=0}^{K-1} g(t_k, x_k^{n', \widehat{\mathcal{Z}}}) - \sum_{m:\tau_m^{n'} < T} \kappa(\tau_m^{n'}, x_{\tau_m^{n'}}^{n', \widehat{\mathcal{Z}}}, z_m^{n'}) + G(x_T^{n', \widehat{\mathcal{Z}}})\right\} \tag{7.12}$$

Algorithm 1 Regression Monte Carlo using GP surrogates for impulse control

---

Require: $K = T/\Delta t$ (time steps), (N_k) (simulation budget per step), w (path lookahead)
1: Set $\widehat{q}(K, \cdot) = G(\cdot)$
2: **for** $k = K - 1, \ldots, 0$ **do**
3:     Generate training design $\mathcal{D}_k \leftarrow (x_k^{(k),1:N_k})$ of size N_k
4:     **for** $n = 1, \ldots, N_k$ **do**
5:         Set $y_{k+1}^n \leftarrow 0$ // pathwise rewards
6:         **for** $\ell = k + 1, \ldots, k + w \wedge K$ **do**
7:             Sample $x_{\ell-1}^{(k),n} \mapsto x_\ell^{(k),n}$ // pathwise controlled trajectories
8:             Set $y_{k+1}^n \leftarrow y_{k+1}^n + g(t_{\ell-1}, x_{\ell-1}^{(k),n})$
9:             Evaluate $q_\ell^{(k),n} = \widehat{q}(\ell, x_\ell^{(k),n})$, $m_\ell^{(k),n} = \widehat{M}(\ell, x_\ell^{(k),n})$ and $\widehat{\mathcal{Z}}_\ell(x_\ell^{(k),n})$
10:            If $m_\ell^{(k),n} > q_\ell^{(k),n}$ set $\begin{cases} x_\ell^{(k),n} \leftarrow x_\ell^{(k),n} + \widehat{\mathcal{Z}}_\ell(x_\ell^{(k),n}) \\ y_{k+1}^n \leftarrow y_{k+1}^n - \kappa(t_\ell, x_\ell^{(k),n}, \widehat{\mathcal{Z}}_\ell(x_\ell^{(k),n})) \end{cases}$ // impulse
11:        **end for**
12:        Set $y_{k+1}^n \leftarrow y_{k+1}^n + \max\left(\widehat{q}(k + w, x_{k+w}^{(k),n}), \widehat{M}(k + w, x_{k+w}^{(k),n})\right)$.
13:    **end for**
14:    Fit the GP $\widehat{q}(t_k, \cdot)$ by regressing $\{y_{k+1}^{1:N_k}\}$ on $\{x_k^{(k),1:N_k}\}$
15: **end for**
16: Return the sequence of GP models $\{\widehat{q}(t_k, \cdot)\}_{k=0}^{K-1}$

---

where $(\tau_m^{n'}, z_m^{n'})$ are the pathwise impulse times and impulse amounts on the n'-path.

The computation of $\widehat{\mathcal{Z}}_k(x)$ is embedded deep in Algorithm 1 and drives the outputs $y_{k+1}^{1:N_k}$ used to fit $\widehat{q}(t_k, \cdot)$. The base implementation is to directly solve (7.10) by calling an optimization sub-routine. The objective function in (7.11) is given *implicitly* in terms of the object $\widehat{q}(t_k, \cdot)$ so a general-purpose, gradient-free optimizer may be needed. Given that $\widehat{M}(t_k, \cdot)$ has to be evaluated repeatedly (line 9 of Algorithm 1) on each forward path emanating from each training input $x^{(k),n}$ this is the major computational bottleneck. To overcome it, several efficiencies could be exploited. First, one may speed up the computation by using a gradient-based optimizer. This requires specifying $\partial_x \widehat{q}(t_k, \cdot)$ in an explicit functional way and is one motivation to use GP surrogates whose gradients are readily available, see Chap. 4.

Second, when computing $\widehat{M}(t_k, x_k^{(k),n})$ for each training input $x_k^{(k),n}, n = 1, \ldots, N_k$, one can record and save the resulting optimal impulse amount z_k^n. Using these training samples, one may then train a secondary GP surrogate $\widetilde{\mathcal{Z}}(t_k, \cdot)$ based on the dataset $(x_k^{(k),1:N_k}, z_k^{1:N_k})$. In subsequent calls the prediction $\widetilde{\mathcal{Z}}(t_k, x')$ is substituted for $\arg\max_z\{\widehat{q}(t_k, x' + z) - \kappa(t_k, x', z)\}$ at subsequent x''s, which is typically much faster. This means that one trains an auxiliary surrogate for the impulse amounts, bypassing the optimization sub-routine.

Third, one may exploit specific features of the problem setting. Consider the common case where the impulse cost $\kappa(t_k, x, z)$ is linear in impulse magnitude z, namely $\kappa(t_k, x, z) = c_0^k z + c_1^k$ for some constants c_0^k, c_1^k. In this situation, the optimization defining $\widehat{M}(t_k, x)$ simplifies considerably. Indeed, the first order conditions for (7.11) reduce to searching for the universal impulse target S_k^* defined by

$$S_k^* = \text{argsup}_z \{\widehat{q}(t_k, x + z) - \kappa(x, z)\} = \text{argsup}_w \{\widehat{q}(t_k, w) - c_0^k(w - x) - c_1^k\}$$
$$\Longleftrightarrow \partial_x \widehat{q}(t_k, S_k^*) = c_0^k. \tag{7.13}$$

It follows that the target level $S_k^* = x + z_k^*(x)$ is independent of the current state x, and moreover can be determined by a single root search on the gradient of the q-function. The structure (7.13)—either do nothing or reset x to level S_k^*—drastically simplifies and stabilizes the numerics, since we just need to determine S_k^* once, and can then immediately evaluate $\widehat{M}(t_k, x) = \widehat{q}(t_k, S_k^*) - c_0 S_k^* + c_0 x - c_1$ for any input x. Consequently, the action region is $\widehat{\mathfrak{S}}_k = \{x : \widehat{q}(t_k, S_k^*) - \widehat{q}(t_k, x) > -c_0^k(x - S_k^*) + c_1^k\}$.

Further Reading

Somewhat overshadowed by the runaway success of deep learning for control, GPs continue to be an attractive tool for implementing dynamic programming from first principles. Indeed, unlike direct empirical risk minimization methods that fully parametrize the policy map and then optimize it via gradient descent, GPDP allows a more interpretable strategy where the algorithm can adaptively (and sequentially) select the training sets and one can inspect and fine-tune the intermediate $\hat{u}(\cdot)$ and $\hat{V}(\cdot)$ functions. Our presentation is based on [10, 31, 112, 115].

A tutorial on GPDP aimed at the control community is by Miao et al. [103].

Appendix A
Mathematical Background

A.1 Matrices and Linear Algebra

We assume the reader is familiar with linear algebra basics. A symmetric matrix is said to be *positive definite (pd)* if for all $\mathbf{z} \in \mathbb{R}^n$, $\mathbf{z}^\top A \mathbf{z} > 0$. If instead $\mathbf{z}^\top A \mathbf{z} \geq 0$, it is *positive semidefinite (psd)*. An alternative characterization is depending on the eigenvalues of a symmetric matrix A. Denoting the ith eigenvalue as λ_i, A is pd if $\lambda_i > 0$ for all i, and psd if $\lambda_i \geq 0$ for all i.

A useful tool is *blocking*, to express the $(n + m) \times (n + m)$ matrix M matrix as

$$M = \begin{bmatrix} A & B \\ C & D \end{bmatrix},$$

where A, B, C, D are of size $n \times n, n \times m, m \times n, m \times m$ respectively. This is particularly useful for matrix inversion:

$$M^{-1} = \begin{bmatrix} A^{-1} + A^{-1}B(D - CA^{-1}B)^{-1}CA^{-1} & -A^{-1}B(D - CA^{-1}B)^{-1} \\ -(D - CA^{-1}B)^{-1}CA^{-1} & (D - CA^{-1}B)^{-1} \end{bmatrix}. \tag{A.1}$$

when all inverses exist.

The Woodbury matrix identity is also useful [84]:

$$(\mathbf{D} + \mathbf{U}\mathbf{B}\mathbf{V})^{-1} - \mathbf{D}^{-1} - \mathbf{D}^{-1}\mathbf{U}(\mathbf{B}^{-1} + \mathbf{V}\mathbf{D}^{-1}\mathbf{U})^{-1}\mathbf{V}\mathbf{D}^{-1} \tag{A.2}$$

$$|\mathbf{D} + \mathbf{U}\mathbf{B}\mathbf{V}| = |\mathbf{B}^{-1} + \mathbf{V}\mathbf{D}^{-1}\mathbf{U}| \times |\mathbf{B}| \times |\mathbf{D}|. \tag{A.3}$$

where $\mathbf{D}$ and $\mathbf{B}$ are invertible matrices of size $N \times N$ and $n \times n$ respectively, and $\mathbf{U}$ and $\mathbf{V}^\top$ are of size $N \times n$. This is commonly used for numerical improvements,

© The Author(s), under exclusive license to Springer Nature Switzerland AG 2025
M. Ludkovski, J. Risk, *Gaussian Process Models for Quantitative Finance*,
SpringerBriefs in Quantitative Finance,
https://doi.org/10.1007/978-3-031-80874-6

e.g. when $\mathbf{D}^{-1}$ is known and $\mathbf{B}$ is of much smaller size $n \ll N$ than $\mathbf{D}^{-1}$. Another tool for numerical improvements is the *Cholesky decomposition*, which involves decomposing $\mathbf{A}$ into a product of the lower triangular matrix $\mathbf{L}$ and its transpose:

$$\mathbf{L}\mathbf{L}^\top = \mathbf{A}, \tag{A.4}$$

where $\mathbf{L}$ is sometimes called the Cholesky factor. This decomposition always exists and is unique for symmetric pd matrices. Its implementation is common and should be expected when linear algebra routines are used. For example, the R package DiceKriging [144] and Python's GPyTorch [65] both use it. This has many benefits including a speed-up of determinant calculations since $|A| = \prod L_{ii}^2$ where L_{ii} is the ith diagonal entry of $\mathbf{L}$.

A.2   Multivariate Normal Distributions

A random vector $\mathbf{Y}$ follows the *multivariate normal distribution (MVN)*, written $\mathbf{Y} \sim \mathcal{MVN}(\boldsymbol{\mu}, \Sigma)$ if its pdf is given by

$$p(\mathbf{y}) = \frac{1}{(2\pi)^{n/2}|\Sigma|^{1/2}} \exp\left(-\frac{1}{2}(\mathbf{y} - \boldsymbol{\mu})^\top \Sigma^{-1}(\mathbf{y} - \boldsymbol{\mu})\right) \tag{A.5}$$

where $\boldsymbol{\mu} = \mathbb{E}[\mathbf{Y}]$ and Σ is the pd covariance matrix of $\mathbf{Y}$. It can be defined in terms of its characteristic function $\exp(i\boldsymbol{\mu}^\top \mathbf{t} - 1/2\mathbf{t}^\top \Sigma \mathbf{t})$. If Σ is psd but not pd, then the pdf of $\mathbf{Y}$ does not exist and is instead defined solely through its characteristic function. A useful property of the MVN is its consistency under marginalization and conditioning.

Proposition A.1 (Marginalization and Conditioning Properties of MVN)
Given a random vector $\mathbf{Y} = [\mathbf{Y}_1, \mathbf{Y}_2]^\top \sim \mathcal{MVN}(\boldsymbol{\mu}, \Sigma)$, *where* $\boldsymbol{\mu} = [\boldsymbol{\mu}_1, \boldsymbol{\mu}_2]^\top$ *and* $\Sigma = \begin{bmatrix} \Sigma_{11} & \Sigma_{12} \\ \Sigma_{21} & \Sigma_{22} \end{bmatrix}$ *in block format, the marginal and conditional distributions are also normal:*

$$\mathbf{Y}_1 \sim \mathcal{N}(\boldsymbol{\mu}_1, \Sigma_{11}), \qquad \mathbf{Y}_2 \sim \mathcal{N}(\boldsymbol{\mu}_2, \Sigma_{22}) \tag{A.6}$$

$$\mathbf{Y}_2|\mathbf{Y}_1 = \mathbf{y}_1 \sim \mathcal{N}\left(\boldsymbol{\mu}_2 + \Sigma_{21}\Sigma_{11}^{-1}(\mathbf{y}_1 - \boldsymbol{\mu}_1), \Sigma_{22} - \Sigma_{21}\Sigma_{11}^{-1}\Sigma_{12}\right). \tag{A.7}$$

Additionally, the MVN works well with affine transformations in the sense that $\mathbf{Z} = A\mathbf{Y} + \mathbf{b}$ satisfies

$$\mathbf{Z} \sim \mathcal{MVN}(A\boldsymbol{\mu} + \mathbf{b}, A\Sigma A^\top). \tag{A.8}$$

A common application is for simulating $\mathbf{Y} \sim \mathcal{MVN}(\boldsymbol{\mu}, \Sigma) \in \mathbb{R}^n$ by generating $\mathbf{U} \sim \mathcal{MVN}(\mathbf{0}, I)$ (n independent standard normals) and computing the Cholesky decomposition of Σ; the resulting $\mathbf{Y} = \boldsymbol{\mu} + \mathbf{LU}$ is $\mathcal{MVN}(\boldsymbol{\mu}, \Sigma)$.

A.3 Differentiability, Smoothness, and Function Spaces

In this section, assume all functions are real valued unless otherwise stated. For $X \subset \mathbb{R}^d$, denote $C^m(X)$ as the space of m-times differentiable functions over X whose derivatives are continuous, $C^\infty(X)$ as the space of infinitely differentiable functions, and $C^0(X)$ is the space of continuous functions. A subscript "0" means compact support, e.g. $C_0^\infty(X)$ is the space of infinitely differentiable functions with compact support in X. The space $L^2(X)$ consists of all square-integrable functions from X to $\mathbb{R}$ (with measure ν understood):

$$\int_X |f(\mathbf{x})|^2 d\nu(\mathbf{x}) < \infty. \tag{A.9}$$

Further, for $0 < \alpha \leq 1$, we say f is α-Hölder continuous on X if there exists a constant K_0 such that

$$|f(\mathbf{x} + \mathbf{h}) - f(\mathbf{h})| \leq K_0 \|\mathbf{h}\|^\alpha, \qquad \mathbf{x}, \mathbf{h} \in X. \tag{A.10}$$

Hölder continuity is a notion of smoothness that is weaker than continuously differentiable but stronger than the typical notion of continuity. The $\alpha = 1$ case is the case of Lipschitz continuity.

A.3.1 Sobolev Spaces

Another notion of smoothness is done through weak derivatives and Sobolev spaces. Consider $L^2(X)$ with $f \in L^2(X)$. A function in $L^2(X)$ is called the *kth weak derivative* of f, denoted $D^k f$, if for every *test function* $\phi \in C_0^\infty(X)$, we have

$$\int_X f(\mathbf{x}) D^k \phi(\mathbf{x}) d\mathbf{x} = (-1)^k \int_X D^k f(\mathbf{x}) \phi(\mathbf{x}) d\mathbf{x}, \tag{A.11}$$

where $D^k \phi(\mathbf{x})$ is the classic derivative of $\phi(\mathbf{x})$, defined through the multi-index $(k_1, \ldots, k_d)$ of nonnegative integers, where $|k| = k_1 + \cdots k_d$ so that $D^k \phi = \frac{\partial^{|k|} \phi}{\partial x_1^{k_1} \cdots \partial x_d^{k_d}}$. In particular, the weak and classical derivatives coincide for differentiable functions. This provides a more general way to define the notion of differentiability through an integration by parts relation, for example for functions that are almost everywhere (but not everywhere) differentiable. Sobolev spaces are functions spaces defined in terms of weak derivatives. In particular, let $m \in \mathbb{N}$ be a nonnegative integer, then the *Sobolev space* with exponent m, denoted $H^m(X)$, is defined as

$$H^m(X) = \left\{ h \in L^2(X) : D^\alpha h \in L^2(X) \text{ for all } \alpha, |\alpha| \leq m \right\} \tag{A.12}$$

equipped with norm

$$\|h\|_m^2 = \sum_{|\alpha| \leq m} \int_X |D^\alpha h(\mathbf{x})|^2 d\mathbf{x}, \tag{A.13}$$

where the sum is over all multi-indices α with $|\alpha| \leq m$. One easier way to better understand $f \in H^m(X)$ is through one of the Sobolev embedding theorems [2].

Proposition A.2 *The function $f \in H^m(X)$ has a unique representer h (meaning $\|f - h\|_m = 0$), which satisfies*

1. *For all k such that $0 \leq k \leq m-1$, the classic derivative $D^k h$ exists, is absolutely continuous and is in $L^2(X)$;*
2. *$D^m h \in L^2(X)$ and is defined almost everywhere.*

Conversely, if f admits a representer satisfying these two conditions, then $f \in H^m(X)$.

This result relates one's understanding of classical derivatives and smoothness "almost everywhere" to Sobolev spaces. Note that we have only defined weak derivatives and Sobolev spaces for integer orders. For non-integer m, one can think of $H^m(X)$ (under certain conditions) as a space which lies "in between" $H^{\lfloor m \rfloor}(X)$ and $H^{\lfloor m \rfloor + 1}$, where $\lfloor m \rfloor$ is the greatest integer less than or equal to m. This can be made formal through interpolation spaces and the interested reader should explore [2] for more details.

References

1. Ackerer, D., Tagasovska, N., Vatter, T.: Deep smoothing of the implied volatility surface. Adv. Neural Inf. Process. Syst. **33**, 11552–11563 (2020)
2. Adams, R.A., Fournier, J.J.: Sobolev Spaces. Elsevier, Amsterdam (2003)
3. Adler, R.J.: The Geometry of Random Fields. Wiley, Hoboken (1981)
4. Agrell, C.: Gaussian processes with linear operator inequality constraints. J. Mach. Learn. Res. **20**(135), 1–36 (2019)
5. Alvarez, M., Lawrence, N.: Sparse convolved Gaussian processes for multi-output regression. Adv. Neural Inf. Process. Syst. **21**, 57–64 (2008)
6. Alvarez, M.A., Rosasco, L., Lawrence, N.D.: Kernels for vector-valued functions: a review. Found. Trends Mach. Learn. **4**(3), 195–266 (2012)
7. Ankenman, B., Nelson, B.L., Staum, J.: Stochastic kriging for simulation metamodeling. Oper. Res. **58**(2), 371–382 (2010)
8. Aronszajn, N.: Theory of reproducing kernels. Trans. Am. Math. Soc. **68**(3), 337–404 (1950)
9. Bachoc, F., López-Lopera, A.F., Roustant, O.: Sequential construction and dimension reduction of Gaussian processes under inequality constraints. SIAM J. Math. Data Sci. **4**(2), 772–800 (2022)
10. Balata, A., Ludkovski, M., Maheshwari, A., Palczewski, J.: Statistical learning for probability-constrained stochastic optimal control. Eur. J. Oper. Res. **290**(2), 640–656 (2021)
11. Basei, M.: Optimal price management in retail energy markets: an impulse control problem with asymptotic estimates. Math. Methods Oper. Rese. **89**(3), 355–383 (2019)
12. Bauer, M., Van der Wilk, M., Rasmussen, C.E.: Understanding probabilistic sparse Gaussian process approximations. Adv. Neural Inf. Process. Syst. **29**, 1533–1541 (2016)
13. Belak, C., Christensen, S., Seifried, F.T.: A general verification result for stochastic impulse control problems. SIAM J. Control Optim. **55**(2), 627–649 (2017)
14. Belomestny, D., Schoenmakers, J.: Advanced Simulation-Based Methods for Optimal Stopping and Control: With Applications in Finance. Springer, Berlin (2018)
15. Bennedsen, M.: Semiparametric estimation and inference on the fractal index of Gaussian and conditionally Gaussian time series data. Econ. Rev. **39**(9), 875–903 (2020)
16. Bennedsen, M., Lunde, A., Pakkanen, M.S.: Decoupling the short-and long-term behavior of stochastic volatility. J. Financ. Econ. **20**(5), 961–1006 (2022)
17. Bensoussan, A., Chevalier-Roignant, B.: Sequential capacity expansion options. Oper. Res. **67**(1), 33–57 (2019)
18. Benth, F.E.: Kriging smooth energy futures curves. Energy Risk, 64–69 (2015)

19. Berlinet, A., Thomas-Agnan, C.: Reproducing Kernel Hilbert Spaces in Probability and Statistics. Springer, Berlin (2011)
20. Bhat, H.S., Kumar, N.: On the Derivation of the Bayesian Information Criterion. School of Natural Sciences, University of California, vol. 99 (2010)
21. Binois, M., Gramacy, R.B.: hetGP: Heteroskedastic Gaussian process modeling and design under replication. R package version 1.0.0 (2017)
22. Binois, M., Gramacy, R.B., Ludkovski, M.: Practical heteroskedastic Gaussian process modeling for large simulation experiments. J. Comput. Graph. Stat. **27**(4), 808–821 (2018)
23. Bonilla, E.V., Chai, K.M., Williams, C.: Multi-task Gaussian process prediction. In: Advances in Neural Information Processing Systems, vol. 20, pp. 153–160 (2008)
24. Cadenillas, A., Zapatero, F.: Classical and impulse stochastic control of the exchange rate using interest rates and reserves. Math. Financ. **10**(2), 141–156 (2000)
25. Camenzind, N., Filipović, D.: Stripping the Swiss discount curve using kernel ridge regression. Eur. Actuarial J. **14**, 371–410 (2024)
26. Capponi, A., Lehalle, C.A.: Machine Learning in Financial Markets: A Guide to Contemporary Practice. Cambridge University Press (2023)
27. Capriotti, L., Jiang, Y., Macrina, A.: AAD and least-square Monte Carlo: Fast Bermudan-style options and XVA Greeks. Algorithm. Financ. **6**(1–2), 35–49 (2017)
28. Carrière, J.F.: Valuation of the early-exercise price for options using simulations and nonparametric regression. Insur. Math. Econ. **19**, 19–30 (1996)
29. Chataigner, M., Cousin, A., Crépey, S., Dixon, M., Gueye, D.: Beyond surrogate modeling: learning the local volatility via shape constraints. SIAM J. Financ. Math. **12**(3), SC58–SC69 (2021)
30. Chen, L., Xu, S.: Deep neural tangent kernel and laplace kernel have the same RKHS. arXiv:2009.10683 (2020)
31. Chen, T., Ludkovski, M.: A machine learning approach to adaptive robust utility maximization and hedging. SIAM J. Financ. Math. **12**(3), 1226–1256 (2021)
32. Cho, Y., Saul, L.K.: Kernel methods for deep learning. In: Advances in Neural Information Processing Systems, vol. 22, pp. 342–350 (2009)
33. Christianson, R.B., Pollyea, R.M., Gramacy, R.B.: Traditional kriging versus modern Gaussian processes for large-scale mining data. Stat. Anal. Data Min. ASA Data Sci. J. **16**(5), 488–506 (2023)
34. Cousin, A., Gueye, D.: Kriging for implied volatility surface. Tech. Rep., HAL Science (2021)
35. Cousin, A., Maatouk, H., Rullière, D.: Kriging of financial term-structures. Eur. J. Oper. Res. **255**(2), 631–648 (2016)
36. Cousin, A., Deleplace, A., Misko, A.: Gaussian process regression for swaption cube construction under no-arbitrage constraints. Risks **10**(12), 232 (2022)
37. Crépey, S., Dixon, M.F.: Gaussian Process regression for derivative portfolio modeling and application to Credit Valuation Adjustment computations. J. Comput. Financ. **24**(1), 47–81 (2020)
38. Cressie, N., Huang, H.C.: Classes of nonseparable, spatio-temporal stationary covariance functions. J. Am. Stat. Assoc. **94**(448), 1330–1339 (1999)
39. Datta, A., Banerjee, S., Finley, A.O., Gelfand, A.E.: Hierarchical nearest-neighbor Gaussian process models for large geostatistical datasets. J. Am. Stat. Assoc. **111**(514), 800–812 (2016)
40. Davis, J., Devos, L., Reyners, S., Schoutens, W.: Gradient boosting for quantitative finance. J. Comput. Financ. **24**(4), 1–40 (2020)
41. De Spiegeleer, J., Madan, D.B., Reyners, S., Schoutens, W.: Machine learning for quantitative finance: fast derivative pricing, hedging and fitting. Quant. Financ. **18**(10), 1635–1643 (2018)
42. de Wolff, T., Cuevas, A., Tobar, F.: MOGPTK: the multi-output Gaussian process toolkit. Neurocomputing **424**, 49–53 (2021)
43. Debón, A., Martínez-Ruiz, F., Montes, F.: A geostatistical approach for dynamic life tables: the effect of mortality on remaining lifetime and annuities. Insur. Math. Econ. **47**(3), 327–336 (2010)

44. Deisenroth, M.P., Rasmussen, C.E., Peters, J.: Gaussian process dynamic programming. Neurocomputing **72**(7), 1508 –1524 (2009)
45. Deschatre, T., Mikael, J.: Deep combinatorial optimisation for optimal stopping time problems: application to swing options pricing. arXiv:2001.11247 (2020)
46. Deville, Y., Ginsbourger, D., Roustant, O., Durande, N.: kergp: Gaussian Process Laboratory. R package version 0.5.8 (2024). https://CRAN.R-project.org/
47. Di Nunno, G., Kubilius, K., Mishura, Y., Yurchenko-Tytarenko, A.: From constant to rough: a survey of continuous volatility modeling. Mathematics **11**(19), 4201 (2023)
48. Dixon, M.F., Halperin, I., Bilokon, P.: Machine Learning in Finance, vol. 1170. Springer, Berlin (2020)
49. Dumas, B., Fleming, J., Whaley, R.E.: Implied volatility functions: empirical tests. J. Financ. **53**(6), 2059–2106 (1998)
50. Duvenaud, D.: Automatic model construction with Gaussian processes. Ph.D. Thesis, University of Cambridge (2014)
51. Duvenaud, D.: The kernel cookbook: advice on covariance functions. Tech. Rep., University of Toronto (2014). https://www.cs.toronto.edu/~duvenaud/cookbook/
52. Duvenaud, D., Lloyd, J., Grosse, R., Tenenbaum, J., Zoubin, G.: Structure discovery in nonparametric regression through compositional kernel search. In: International Conference on Machine Learning, pp. 1166–1174. PMLR (2013)
53. Egloff, D.: Monte Carlo algorithms for optimal stopping and statistical learning. Ann. Appl. Probab. **15**(2), 1396–1432 (2005)
54. Egloff, D., Kohler, M., Todorovic, N.: A dynamic look-ahead Monte Carlo algorithm for pricing Bermudan options. Ann. Appl. Probab. **17**(4), 1138–1171 (2007)
55. Erickson, C., Ankenman, B.E., Sanchez, S.M.: Comparison of Gaussian process modeling software. In: Winter Simulation Conference (WSC), 2016, pp. 3692–3693. IEEE, Piscataway (2016)
56. Federico, S., Rosestolato, M., Tacconi, E.: Irreversible investment with fixed adjustment costs: a stochastic impulse control approach. Math. Financ. Econ. **13**(4), 579–616 (2019)
57. Ferguson, R., Green, A.D.: Deeply learning derivatives. SSRN 3244821 (2018)
58. Filipović, D., Pasricha, P.: Empirical asset pricing via ensemble Gaussian Process regression. arXiv:2212.01048 (2022)
59. Filipović, D., Pelger, M., Ye, Y.: Stripping the discount curve-a robust machine learning approach. Tech. Rep. 22–24, Swiss Finance Institute Research Paper (2022)
60. Flaxman, S., Gelman, A., Neill, D., Smola, A., Vehtari, A., Wilson, A.G.: Fast hierarchical Gaussian processes. Working paper (2015)
61. Fu, H., Jin, X., Pan, G., Yang, Y.: Estimating multiple option Greeks simultaneously using random parameter regression. J. Comput. Financ. **16**(2), 85 (2012)
62. Gan, G., Valdez, E.A.: Valuation of large variable annuity portfolios: Monte Carlo simulation and synthetic datasets. Depend. Model. **5**(1), 354–374 (2017)
63. Gan, G., Valdez, E.A.: Metamodeling for Variable Annuities. CRC Press, Boca Raton (2019)
64. Garcia, R., Gençay, R.: Pricing and hedging derivative securities with neural networks and a homogeneity hint. J. Econ. **94**(1–2), 93–115 (2000)
65. Gardner, J.R., Pleiss, G., Bindel, D., Weinberger, K.Q., Wilson, A.G.: GPyTorch: blackbox matrix-matrix Gaussian process inference with GPU acceleration. In: Proceedings of the 32nd International Conference on Neural Information Processing Systems, pp. 7587–7597 (2018)
66. Garrido-Merchán, E.C., Hernández-Lobato, D.: Dealing with categorical and integer-valued variables in Bayesian optimization with Gaussian processes. Neurocomputing **380**, 20–35 (2020)
67. Gatheral, J.: The Volatility Surface: A Practitioner's Guide. Wiley, Hoboken (2011)
68. Genton, M.G.: Classes of kernels for machine learning: a statistics perspective. J. Mach. Learn. Res. **2**, 299–312 (2001)
69. Gneiting, T.: Criteria of pólya type for radial positive definite functions. Proc. Am. Math. Soc. **129**(8), 2309–2318 (2001)
70. Gneiting, T.: Compactly supported correlation functions. J. Mult. Anal. **83**(2), 493–508 (2002)

71. Gneiting, T.: Nonseparable, stationary covariance functions for space–time data. J. Am. Stat. Assoc. **97**(458), 590–600 (2002)
72. Gneiting, T., Schlather, M.: Stochastic models that separate fractal dimension and the hurst effect. SIAM Rev. **46**(2), 269–282 (2004)
73. Golchi, S., Bingham, D., Chipman, H., Campbell, D.: Monotone emulation of computer experiments. SIAM/ASA J. Uncertain. Quant. **3**(1), 370–392 (2015)
74. Goldberg, P.W., Williams, C.K., Bishop, C.M.: Regression with input-dependent noise: a Gaussian process treatment. In: Advances in Neural Information Processing Systems, vol. 10, pp. 493–499. MIT Press, Cambridge (1998)
75. Goudenège, L., Molent, A., Zanette, A.: Machine learning for pricing American options in high-dimensional Markovian and non-Markovian models. Quant. Financ. **20**(4), 573–591 (2020)
76. Goudenège, L., Molent, A., Zanette, A.: Gaussian process regression for pricing variable annuities with stochastic volatility and interest rate. Decis. Econ. Financ. **44**, 57–72 (2021)
77. Gramacy, R.B.: tgp: an R package for Bayesian nonstationary, semiparametric nonlinear regression and design by treed Gaussian process models. J. Stat. Softw. **19**(9), 6 (2007)
78. Gramacy, R.B.: laGP: large-scale spatial modeling via local approximate Gaussian processes in R. J. Stat. Softw. **72**, 1–46 (2016)
79. Gramacy, R.B.: Surrogates: Gaussian Process Modeling, Design, and Optimization for the Applied Sciences. Chapman and Hall/CRC, Boca Raton (2020)
80. Gramacy, R.B., Apley, D.W.: Local Gaussian process approximation for large computer experiments. J. Comput. Graph. Stat. **24**(2), 561–578 (2015)
81. Gramacy, R.B., Lee, H.K.H.: Bayesian treed Gaussian process models with an application to computer modeling. J. Am. Stat. Assoc. **103**(483), 1119–1130 (2008)
82. Gramacy, R.B., Ludkovski, M.: Sequential design for optimal stopping problems. SIAM J. Financ. Math. **6**(1), 748–775 (2015)
83. Hartikainen, J., Särkkä, S.: Kalman filtering and smoothing solutions to temporal Gaussian process regression models. In: 2010 IEEE International Workshop on Machine Learning for Signal Processing, pp. 379–384. IEEE, Piscataway (2010)
84. Harville, D.: Matrix Algebra from a Statistician's Perspective. Springer, Berlin (1997)
85. Hu, R., Ludkovski, M.: Sequential design for ranking response surfaces. SIAM/ASA J. Uncertain. Quant. **5**(1), 212–239 (2017)
86. Huynh, N., Ludkovski, M.: Multi-output Gaussian processes for multi-population longevity modelling. Ann. Actuarial Sci. **15**(2), 318–345 (2021)
87. Huynh, N., Ludkovski, M.: Joint models for cause-of-death mortality in multiple populations. Ann. Actuarial Sci. **18**(1), 51–77 (2024)
88. Jähnichen, P., Wenzel, F., Kloft, M., Mandt, S.: Scalable generalized dynamic topic models. In: Proceedings of the Twenty-First International Conference on Artificial Intelligence and Statistics (AISTATS), vol. 84, pp. 1427–1435. PMLR (2018)
89. Jeffreys, H.: The Theory of Probability. Oxford University Press, Oxford (1961)
90. Jylänki, P., Vanhatalo, J., Vehtari, A.: Robust Gaussian process regression with a Student-t likelihood. J. Mach. Learn. Res. **12**, 3227–3257 (2011)
91. Kanagawa, M., Hennig, P., Sejdinovic, D., Sriperumbudur, B.K.: Gaussian processes and kernel methods: a review on connections and equivalences. arXiv:1807.02582 (2018)
92. Katzfuss, M., Guinness, J.: A general framework for Vecchia approximations of Gaussian processes. Stat. Sci. **36**(1), 124–141 (2021)
93. Katzfuss, M., Guinness, J., Gong, W., Zilber, D.: Vecchia approximations of Gaussian-process predictions. J. Agr. Biol. Environ. Stat. **25**(3), 383–414 (2020)
94. Kaufman, C.G., Bingham, D., Habib, S., Heitmann, K., Frieman, J.A.: Efficient emulators of computer experiments using compactly supported correlation functions, with an application to cosmology. Ann. Appl. Stat. **5**(4), 2470–2492 (2011)
95. Kharroubi, I., Langrené, N., Pham, H.: A numerical algorithm for fully nonlinear HJB equations: an approach by control randomization. Monte Carlo Methods Appl. **20**(2), 145–165 (2014)

96. Kim, H.M., Mallick, B.K., Holmes, C.C.: Analyzing nonstationary spatial data using piecewise Gaussian processes. J. Am. Stat. Assoc. **100**(470), 653–668 (2005)

97. Kohler, M.: A regression-based smoothing spline Monte Carlo algorithm for pricing American options in discrete time. Adv. Stat. Anal. **92**(2), 153–178 (2008)

98. Kohler, M., Krzyżak, A., Todorovic, N.: Pricing of high-dimensional American options by neural networks. Math. Financ. **20**(3), 383–410 (2010)

99. Kostantinos, N.: Gaussian mixtures and their applications to signal processing. In: Advanced Signal Processing Handbook: Theory and Implementation for Radar, Sonar, and Medical Imaging Real Time Systems, pp. 1–3. CRC Press, Boca Raton (2000)

100. Lalchand, V., Rasmussen, C.E.: Approximate inference for fully Bayesian Gaussian process regression. In: Symposium on Advances in Approximate Bayesian Inference, pp. 1–12. PMLR (2020)

101. Laurini, M.P., Ohashi, A.: A noisy principal component analysis for forward rate curves. Eur. J. Oper. Res. **246**(1), 140–153 (2015)

102. Lin, Q., Hu, J., Zhou, Q., Cheng, Y., Hu, Z., Couckuyt, I., Dhaene, T.: Multi-output Gaussian process prediction for computationally expensive problems with multiple levels of fidelity. Knowl. Based Syst. **227**, 107151 (2021)

103. Liu, M., Chowdhary, G., Da Silva, B.C., Liu, S.Y., How, J.P.: Gaussian processes for learning and control: a tutorial with examples. IEEE Control Syst. Mag. **38**(5), 53–86 (2018)

104. Liu, S., Borovykh, A., Grzelak, L.A., Oosterlee, C.W.: A neural network-based framework for financial model calibration. J. Math. Ind. **9**(1), 1–28 (2019)

105. Liu, H., Ong, Y.S., Shen, X., Cai, J.: When Gaussian process meets big data: a review of scalable GPs. IEEE Trans. Neural Netw. Learn. Syst. **31**(11), 4405–4423 (2020)

106. Liu, H., Ding, J., Xie, X., Jiang, X., Zhao, Y., Wang, X.: Scalable multi-task Gaussian processes with neural embedding of coregionalization. Knowl. Based Syst. **247**, 108775 (2022)

107. Longstaff, F., Schwartz, E.: Valuing American options by simulations: a simple least squares approach. Rev. Financ. Stud. **14**, 113–148 (2001)

108. López-Lopera, A.F., Bachoc, F., Durrande, N., Roustant, O.: Finite-dimensional Gaussian approximation with linear inequality constraints. SIAM/ASA J. Uncertain. Quant. **6**(3), 1224–1255 (2018)

109. López-Lopera, A., Bachoc, F., Roustant, O.: High-dimensional additive Gaussian processes under monotonicity constraints. Adv. Neural Inf. Process. Syst. **35**, 8041–8053 (2022)

110. Lucia, J.J., Schwartz, E.S.: Electricity prices and power derivatives: evidence from the Nordic power exchange. Rev. Derivatives Res. **5**, 5–50 (2002)

111. Ludkovski, M.: Kriging metamodels and experimental design for Bermudan option pricing. J. Comput. Financ. **22**(1), 37–77 (2018)

112. Ludkovski, M.: Regression Monte Carlo for impulse control. Math. Action **11**(1), 73–90 (2022)

113. Ludkovski, M.: mlOSP: towards a unified implementation of regression Monte Carlo algorithms. J. Comput. Financ. **17**(1), 59–109 (2023)

114. Ludkovski, M.: Statistical machine learning for quantitative finance. Ann. Rev. Stat. Appl. **10**, 271–295 (2023)

115. Ludkovski, M., Maheshwari, A.: Simulation methods for stochastic storage problems: a statistical learning perspective. Energy Syst. **11**(2), 377–415 (2020)

116. Ludkovski, M., Padilla, D.: Analyzing state-level longevity trends with the US mortality database. arXiv:2312.01518 (2023)

117. Ludkovski, M., Risk, J.: Expressive mortality models through Gaussian Process kernels. ASTIN Bullet. J. IAA **54**(2), 327–359 (2024)

118. Ludkovski, M., Saporito, Y.: KrigHedge: Gaussian process surrogates for delta hedging. Appl. Math. Financ. **28**(4), 330–360 (2021)

119. Ludkovski, M., Risk, J., Zail, H.: Gaussian process models for mortality rates and improvement factors. ASTIN Bullet. J. IAA **48**(3), 1307–1347 (2018)

120. Lyu, X., Ludkovski, M.: Adaptive batching for Gaussian process surrogates with application in noisy level set estimation. Stat. Anal. Data Min. ASA Data Sci. J. **15**(2), 225–246 (2022)
121. Maatouk, H., Bay, X.: Gaussian process emulators for computer experiments with inequality constraints. arXiv:1606.01265 (2016)
122. Maddox, W.J., Stanton, S., Wilson, A.G.: Conditioning sparse variational Gaussian processes for online decision-making. Adv. Neural Inf. Process. Syst. **34**, 6365–6379 (2021)
123. Malkomes, G., Schaff, C., Garnett, R.: Bayesian optimization for automated model selection. Adv. Neural Inf. Process. Syst. **29**, 2900–2908 (2016)
124. Mandelbrot, B.B., Van Ness, J.W.: Fractional Brownian motions, fractional noises and applications. SIAM Rev. **10**(4), 422–437 (1968)
125. Maran, A., Pallavicini, A.: Interpolating commodity futures prices with Kriging. arXiv:2110.13021 (2021)
126. Matérn, B.: Spatial Variation, vol. 36. Springer, Berlin (2013)
127. Max Planck Institute for Demographic Research (Germany) and University of California, Berkeley (USA) and French Institute for Demographic Studies (France): HMD. Human Mortality Database. www.mortality.org
128. McKay, M., Beckman, R., Conover, W.: Comparison of three methods for selecting values of input variables in the analysis of output from a computer code. Technometrics **21**(2), 239–245 (1979)
129. Melsnes, J.E.: Kriging the power futures curve and pricing of options. Master's thesis, University of Oslo (2019)
130. Micchelli, C.A., Xu, Y., Zhang, H.: Universal kernels. J. Mach. Learn. Res. **7**(12), 2651–2667 (2006)
131. Mishura, Y.: Stochastic Calculus for Fractional Brownian Motion and Related Processes. Springer, Berlin (2008)
132. Mishura, Y., Shevchenko, G., Shklyar, S.: Gaussian processes with Volterra kernels. In: International Conference on Stochastic Processes and Algebraic Structures, pp. 249–276. Springer, Berlin (2019)
133. Neal, R.M.: Bayesian Learning for Neural Networks, vol. 118. Springer, Berlin (2012)
134. Parzen, E.: An approach to time series analysis. Ann. Math. Stat. **32**(4), 951–989 (1961)
135. Perdikaris, P., Venturi, D., Royset, J.O., Karniadakis, G.E.: Multi-fidelity modelling via recursive co-kriging and Gaussian–Markov random fields. Proc. R. Soc. A Math. Phys. Eng. Sci. **471**(2179), 20150018 (2015)
136. Porcu, E., Furrer, R., Nychka, D.: 30 Years of space–time covariance functions. Wiley Interdiscip. Rev. Comput. Stat. **13**(2), e1512 (2021)
137. Porcu, E., Bevilacqua, M., Schaback, R., Oates, C.J.: The Matérn model: a journey through statistics, numerical analysis and machine learning. Stat. Sci. **39**(3), 469–492 (2024)
138. Qin, Z., Almeida, C.: A Bayesian nonparametric approach to option pricing. Brazil. Rev. Financ. **18**(4), 115–137 (2020)
139. Rasmussen, C.E., Nickisch, H.: Gaussian processes for machine learning (GPML) toolbox. J. Mach. Learn. Res. **11**, 3011–3015 (2010)
140. Rasmussen, C.E., Williams, C.K.I.: Gaussian Processes for Machine Learning. The MIT Press, Cambridge (2006)
141. Revuz, D., Yor, M.: Continuous Martingales and Brownian Motion, vol. 293. Springer, Berlin (2013)
142. Riihimäki, J., Vehtari, A.: Gaussian processes with monotonicity information. In: Proceedings of the Thirteenth International Conference on Artificial Intelligence and Statistics (AISTATS), pp. 645–652 (2010)
143. Risk, J., Ludkovski, M.: Sequential design and spatial modeling for portfolio tail risk measurement. SIAM J. Financ. Math. **9**(4), 1137–1174 (2018)
144. Roustant, O., Ginsbourger, D., Deville, Y.: Dicekriging, DiceOptim: two R packages for the analysis of computer experiments by kriging-based metamodeling and optimization. J. Stat. Softw. **51**(1), 1–55 (2012)

145. Ruder, S.: An overview of gradient descent optimization algorithms. arXiv:1609.04747 (2016)
146. Ruf, J., Wang, W.: Hedging with linear regressions and neural networks. J. Bus. Econ. Stat. **40**(4), 1442–1454 (2022)
147. Särkkä, S., Solin, A.: Applied Stochastic Differential Equations, vol. 10. Cambridge University Press, Cambridge (2019)
148. Schölkopf, B., Smola, A.J.: Learning with kernels: support vector machines, regularization, optimization, and beyond. MIT Press, Cambridge (2002)
149. Shah, A., Wilson, A., Ghahramani, Z.: Student-t processes as alternatives to Gaussian processes. In: Proceedings of the Seventeenth International Conference on Artificial Intelligence and Statistics, vol. 33, pp. 877–885. PMLR (2014)
150. Shawe-Taylor, J., Cristianini, N.: Kernel Methods for Pattern Analysis. Cambridge University Press, Cambridge (2004)
151. Silverman, R.: Locally stationary random processes. IRE Trans. Inf. Theory **3**(3), 182–187 (1957)
152. Silverman, R.A.: A matching theorem for locally stationary random processes. Commun. Pure Appl. Math. **12**(2), 373–383 (1959)
153. Snelson, E., Ghahramani, Z.: Sparse Gaussian processes using pseudo-inputs. In: Advances in Neural Information Processing Systems, pp. 1257–1264 (2005)
154. Snoek, J., Larochelle, H., Adams, R.P.: Practical Bayesian optimization of machine learning algorithms. Adv. Neural Inf. Process. Syst. **25**, 2951–2959 (2012)
155. Snoek, J., Swersky, K., Zemel, R., Adams, R.: Input warping for Bayesian optimization of non-stationary functions. In: International Conference on Machine Learning, pp. 1674–1682. PMLR (2014)
156. Stein, M.L.: Interpolation of Spatial Data: Some Theory for Kriging. Springer, Berlin (1999)
157. Stein, M.L.: Space–time covariance functions. J. Am. Stat. Assoc. **100**(469), 310–321 (2005)
158. Swiler, L.P., Gulian, M., Frankel, A.L., Safta, C., Jakeman, J.D.: A survey of constrained Gaussian process regression: approaches and implementation challenges. J. Mach. Learn. Model. Comput. **1**(2), 119–156 (2020)
159. Tegnér, M., Roberts, S.: A probabilistic approach to nonparametric local volatility. arXiv:1901.06021 (2019)
160. Titsias, M.: Variational learning of inducing variables in sparse Gaussian processes. In: Proceedings of the Twelfth International Conference on Artificial Intelligence and Statistics (AISTATS), vol. 5, pp. 567–574. PMLR (2009)
161. Tsitsiklis, J.N., van Roy, B.: Regression methods for pricing complex American-style options. IEEE Trans. Neural Netw. **12**(4), 694–703 (2001)
162. Van Der Vaart, A., Van Zanten, H.: Information rates of nonparametric Gaussian process methods. J. Mach. Learn. Res. **12**(6), 2095–2119 (2011)
163. Vanhatalo, J., Jylänki, P., Vehtari, A.: Gaussian process regression with Student-t likelihood. Adv. Neural Inf. Process. Syst. **22**, 1910–1918 (2009)
164. Vanhatalo, J., Riihimäki, J., Hartikainen, J., Jylänki, P., Tolvanen, V., Vehtari, A.: Bayesian modeling with Gaussian processes using the GPstuff toolbox. preprint arXiv:1206.5754 (2012)
165. Wang, W.: On the inference of applying Gaussian process modeling to a deterministic function. Electron. J. Stat. **15**(2), 5014–5066 (2021)
166. Wang, X., Berger, J.O.: Estimating shape constrained functions using Gaussian processes. SIAM/ASA J. Uncertain. Quant. **4**(1), 1–25 (2016)
167. Wikle, C.K., Zammit-Mangion, A., Cressie, N.: Spatio-Temporal Statistics with R. CRC Press, Boca Raton (2019)
168. Wilson, A., Adams, R.: Gaussian process kernels for pattern discovery and extrapolation. In: International Conference on Machine Learning, pp. 1067–1075. PMLR (2013)
169. Wilson, A., Nickisch, H.: Kernel interpolation for scalable structured Gaussian processes (KISS-GP). In: International Conference on Machine Learning, pp. 1775–1784 (2015)

170. Wilson, A.G., Gilboa, E., Nehorai, A., Cunningham, J.P.: GPatt: fast multidimensional pattern extrapolation with Gaussian processes. arXiv:1310.5288 (2013)
171. Yang, X., Barajas-Solano, D., Tartakovsky, G., Tartakovsky, A.M.: Physics-informed CoKriging: a Gaussian process regression-based multifidelity method for data-model convergence. J. Comput. Phys. **395**, 410–431 (2019)

Index

© The Author(s), under exclusive license to Springer Nature Switzerland AG 2025
M. Ludkovski, J. Risk, *Gaussian Process Models for Quantitative Finance*,
SpringerBriefs in Quantitative Finance,
https://doi.org/10.1007/978-3-031-80874-6